Staging Collaborative Design and Innovation

An Action-Oriented Participatory Approach

Edited by

Christian Clausen

Professor Emeritus of Staging and Shaping Design and Innovation, Aalborg University, Denmark

Dominique Vinck

Professor of Science and Technology Studies, University of Lausanne, Switzerland

Signe Pedersen

Assistant Professor of Participatory Design and Engineering Design, Aalborg University, Denmark

Jens Dorland

Assistant Professor of Technological and Socio-economic Planning, Roskilde University, Denmark

Cheltenham, UK • Northampton, MA, USA

Published by
Edward Elgar Publishing Limited
The Lypiatts
15 Lansdown Road
Cheltenham
Glos GL50 2JA
UK

Edward Elgar Publishing, Inc.
William Pratt House
9 Dewey Court
Northampton
Massachusetts 01060
USA

A catalogue record for this book
is available from the British Library

Library of Congress Control Number: 2020948652

This book is available electronically in the **Elgar**online
Social and Political Science subject collection
http://dx.doi.org/10.4337/9781839103438

ISBN 978 1 83910 342 1 (cased)
ISBN 978 1 83910 343 8 (eBook)

Printed and bound by CPI Group (UK) Ltd, Croydon, CR0 4YY

Contents

Figures

Tables

Contributors

Charles Anthony Bates is the Global Head of BU Motors Technology in the Work Functions Division at Danfoss Power Solutions (DPS), where he has worked with the development of new component, material, product and process technologies associated with the design, verification and production of hydraulic motors and steering units since 2005. His Industrial PhD is a collaboration between DPS and Aalborg University Copenhagen and was made possible with funding from Innovation Fund Denmark. In addition to his social science research, he has (co)authored multiple patents and journal articles in the fields of hydraulics and tribology.

Søsser Brodersen is Associate Professor in participatory design and engineering design at the Department of Planning, Aalborg University. Currently she is involved in a research project named PACE focusing on exploring how sensory technologies can contribute to reducing hospitalization of elderly people with dementia. Her research interest lies within how to stage participatory design processes, especially within healthcare products or systems.

Louise Brønnum is the Innovation Process Lead, Innovation Hub at Nets group, a European payment service provider, where she works with early stage innovation and concept development. Louise is a trained design engineer and holds a PhD in 'Strategic Enactment of Front End Innovation: A Case study of multiple enabling opportunities' from 2017. Since finishing her PhD she has been working in the financial industry developing and applying ways of working within early stage innovation at both a strategic and pragmatic level.

Christian Clausen is Professor Emeritus at the section for Sustainable Design and Transition, Department of Planning, Aalborg University Copenhagen. His current research and teaching deals with the staging of design and innovation and he combines an engineering background with insight in Science and Technology Studies (STS) and organization. He has published widely on social shaping of technology and sociotechnical dimensions of design and innovation. A special interest has been concerned with worker and user participation and the management of technological and organizational processes in design and innovation. Recent research projects include staging of innovative processes across knowledge domains and sociotechnical dimensions of the work with product ideas in organizations.

Jens Dorland has been Assistant Professor at Roskilde University, Department of People and Technology, since 2020. He holds a PhD in Transformative Social Innovation from Aalborg University and an MSc in Engineering from the Technical University of Denmark. Jens's research focuses on giving action-oriented insight through his studies in design and innovation in relation to organizational change and sustainable transitions. His approach is very interdisciplinary with a foundation in engineering design and Science and Technology Studies (STS) that has been expanded with an organizational theory perspective, and his current position with a research focus on Technological and Socio-economic Planning.

Eva Guldmann is founder and CEO of Rotundo, a sustainability-oriented management consultancy. She holds a PhD in Sustainable Business Development from Aalborg University and a MSc in Engineering from the Technical University of Denmark. Eva has years of experience working in the manufacturing industry and is passionate about facilitating the implementation of viable circular business models at company level. To this end she does research and hands-on work to support business experimentation, circular business model innovation and companies' overall sustainability journey.

Wendy Gunn (2019-ongoing) is Adjunct Professor (Research) in Design, Emerging Technologies Lab, Faculty of Art, Design and Architecture, Monash University, Australia. During 2019–2020, Gunn was Jinshan Distinguished Professor, School of Arts, Jiangsu University, Zhenjiang, Jiangsu, P.R. China. Gunn's research profile is in the field of Design Anthropology, which is evident in her publication track record, research grants and teaching experience. Gunn has generated important insights into how collaborative processes unfold and take form, particularly in the field of design and innovation, with specific focus on the intersections with anthropology, architecture and design. Gunn's current research, Future Textiles for Healthcare Facilities, builds upon interdisciplinary research on improving air quality in hospitals (2017–2018) and indoor climate and quality of life (2008–2011).

Birgitte Hoffmann is Associate Professor at the section for Sustainable Design and Transition, Aalborg University Copenhagen. Trained as an engineer and urban planner, Birgitte has always worked to bridge the development of large urban infrastructures and urban life. As an action researcher, and taking sustainable transition as the key challenge, Birgitte's research focuses on staging strategic arenas in which various actors and actants can meet to explore and rehearse alternative urban futures. In a series of research projects, based on municipal planners as facilitators, she has published on how to navigate between existing realities and alternative sustainable futures.

Rikke Dorothea Huulgaard (MSc, PhD) has been a Postdoc at Aalborg University, Department of Planning, Denmark since 2016. Her main research areas are circular economy, ecodesign and European Union environmental product policies covering electrical and electronic products. More specifically concerning circular economy, she has collaborated with different companies on how to work with and integrate circular economy into their business models through the research projects Business Strategies for Sustainable Production 3.0, and Circular Innovation in Partnerships.

Charlotte Louise Jensen is a trained researcher in sustainable change and social practice. She holds a PhD in sociology of technology, consumption and transition from Aalborg University. Her research deals with sustainable consumption and how energy is used in different contexts and practices. She focuses particularly on how notions of 'the good life' are experienced as well as practiced and how they are important in shaping consumption dynamics. She has published widely on (energy) consumption, practices and problem framings in energy policy and objectives. She is currently working as an independent researcher and writer.

Ask Greve Johansen has a master's degree in Sociology and is currently finishing his PhD dissertation at Aalborg University Copenhagen on planning practice and how local governments anticipate, plan and govern in the context of global societal challenges such as climate change and biodiversity loss. In his work he pays particular attention to how the notion of value is defined, negotiated and tinkered with in everyday practice. This practice relies on a range of tangible tools such as reports, tables, spreadsheets and, in some cases, socio-economic calculations, forecasts and rankings.

Joakim Juhl is Assistant Professor at the section for Sustainable Design and Transition, Aalborg University Copenhagen. Joakim's background includes engineering and Science and Technology Studies, which he applied in his dissertation on mathematical modelling. His research focuses on knowledge production and innovation, involves both historical and ethnographic approaches and treats epistemological and political questions in relation to how innovation science is understood and practiced in different settings. Joakim is a junior council member of the Science and Democracy Network and has previously worked at the Science, Technology and Society program at Harvard University.

Søren Kerndrup has been Associate Professor at Aalborg University, Department of Planning since 2002. His main research area is innovation and entrepreneurship in projects, network and communities. In recent years, special attention has been paid to the importance of designing spaces and networks that make it possible to challenge traditional linear thinking in companies and networks through the co-creation and co-design of circular business models and activities.

Hanne Lindegaard is Associate Professor at the section for Sustainable Design and Transition, Aalborg University Copenhagen. Hanne has a master's degree in European Ethnology from the University of Copenhagen, and a PhD from the Technical University of Denmark. The combination of the two disciplines has given her strong cross-disciplinary experiences combining ethnographic field study methodology with engineering design approaches. Hanne has worked with participatory design, visualization tools and research on how to motivate multiple stakeholders in co-designing activities.

Peter Munthe-Kaas is a lecturer and researcher connected to the section for Sustainable Design and Transition, Aalborg University Copenhagen. He is interested in imaginaries of urban futures and how these can be brought about through participatory design and experimental urban development. Peter often finds himself working in the interplay between research, urban planning, performance art and activism and has recently founded the consultancy Nyforbyelse, which supports municipalities in moving towards more sustainable and holistic urban development.

Signe Pedersen is Assistant Professor in participatory design and engineering design at the Department of Planning, Aalborg University Copenhagen. Signe has a MSc in Design & Innovation from the Technical University of Denmark and holds a PhD in co-design from Aalborg University. Her research focuses on the staging of active, participatory involvement of multiple and diverse actors in engineering design processes with a special interest in designing for healthcare. Furthermore, Signe has practical and professional experience from working together with public as well as private organizations (from multi-national enterprises to local non-government organizations (NGOs)) in designing innovative solutions together with a multiplicity of users, experts and policy makers etc.

Elizabeth B.-N. Sanders is the founder of MakeTools, LLC where she explores new spaces in the emerging design landscapes. As a practitioner, she introduced many of the methods being used today to drive design from a human-centered perspective. Her current research focuses on co-design processes for health and wellbeing. She joined the Design Department at Ohio State University (OSU) in 2011 after having worked as a design research consultant in industry since 1981. At OSU she invites students to use co-designing to address the significant social, cultural and environmental challenges we face today. She has a PhD in Experimental and Quantitative Psychology and a BA in both Psychology and Anthropology.

Mylène Tanferri is a PhD student in Science and Technology Studies at the University of Lausanne and Information Science at the Federal University of Bahia (Brazil). Her research focuses on visual practices in heritage numerization and looks at how people frame their perceptions and environments to reach agreement about quality. She is currently working on the SFNS (Swiss National Research Foundation) project 'Patrimonialisation du direct' to analyse video data from the Fête des Vignerons.

Dominique Vinck is Professor of Science and Technology Studies at the University of Lausanne and at the College of Humanities of the Swiss Federal Institute of Technology in Lausanne. A member of the STSLab, he directs the University of Lausanne Institute of Social Sciences and the Doctoral Program in Digital Studies. His research focuses on the sociology of science and innovation, mainly in the field of the engineering of digital cultures and humanities. He is Chief Editor of *the Revue d'Anthropologie des Connaissances* and member of the editorial board of *Engineering Studies*. He has published among others: *Everyday Engineering: Ethnography of Design and innovation* (MIT Press, 2003), *The Sociology of Scientific Work: The Fundamental Relationship between Science and Society* (Edward Elgar, 2010), *Critical Studies of Innovation: Alternatives to the Pro-Innovation Bias* (Edward Elgar, 2017) and *Les Métiers de l'Ombre de la Fête des Vignerons* (Antipodes, 2019).

Yutaka Yoshinaka is Associate Professor of Sociotechnical Competences and Innovative Processes at the Technical University of Denmark, near Copenhagen. Yutaka holds a PhD in Science and Technology Studies (STS) and a MSc(Eng.) with a profile in Planning and Technology Management. Recent research deals with new conceptions in addressing design synthesis in innovation processes. Socio-material design and generative design methodologies form the crux of his inquiry at the 'front-end' (or early stages) of design and innovation. Moreover, his research into innovation is informed by a multi-actor orientation toward design-in-use and domestication aspects of technologies, across several empirical domains.

Preface

This book contributes a novel action-oriented, collaborative, and participatory perspective on design and innovation. Based on an exploration of the notion of staging inspired by a theatrical metaphor, it offers a number of in-depth case studies on how professionals practice staging processes for design and innovation.

The idea behind the book takes its main departure from attempts over more than 30 years to inform engineering curricula with sociotechnical dimensions in a Danish context at the Technical University of Denmark and Aalborg University. A number of the contributors to the book have been engaged in the development of the new engineering education programmes in design and innovation/sustainable design that have emerged since the start of the century. These programmes aimed at the creation of a new professional role for engineers enabling them to analyse and handle complex and wicked problems in close collaboration with relevant stakeholders in practice and to take on more social responsibility. The means to do this was to combine and integrate three competence areas: engineering and technical competences, creative design skills, and sociotechnical competences. Collaborative innovative practices, the focus on engineering and materiality and the integration of design skills in organisational practice were considered key challenges for innovation within industry as well as in the public domain.

A strong sense of a new professional identity grew from these activities emphasising university–industry collaboration and the staging and navigation of design and innovation processes. Here, the staging approach eventually became a way to enable these new engineering and similar professions to integrate sociotechnical understandings and collaborative concerns in their work with design and innovation. Many of the authors in this book have either served as teachers in these programmes and/or they have graduated from them and undertaken their postgraduate work in associated research programmes.

Staging as a research theme was developed with strong inspiration from and collaboration with the Scandinavian tradition of participatory design, and attempts to develop Science and Technology Studies (STS) including the shaping of technology into something actionable. Interestingly, staging design and innovation as a theme and the use of a theatrical metaphor seems to be able to attract attention at STS events as well as more engineering design and innovation management focused events.

With this background the book brings together theoretically informed reflection and practical experience from solving real life problems in design and innovation. The close connection between theory and practice is reflected in the authorships in several ways, either by university researchers collaborating closely with industry and public authorities or by professionals undertaking academic research while working in the industry.

The foundation for the book was laid by a number of events hosted by the research group Sustainable Design and Transition at the Department of Planning at Aalborg University in Copenhagen where potential contributors were invited to share and discuss their approaches to staging design and innovation and submit their abstracts for potential contributions. As a next step the international network of collaboration was mobilized in order to create and make visible both theoretical inspirations and practices in staging design and innovation. Consequently, the collection of chapters assembled is based on a coming together of research contributions from a Scandinavian action-oriented tradition with an international outreach to research environments in STS, Design Anthropology and Co-design from Europe, Australia, China and the United States. Also worth mentioning is the European Union (EU) COST research network dating back to the 1990s, gathering technology studies and technology management around a common interest in the role of design in the shaping of technology, shared by Christian Clausen and Dominique Vinck. This book has in particular benefitted from such international synergies in the application of actor–network theory, and in understanding the role of intermediary objects and the theatrical metaphor in the development of an actionable staging approach to design and innovation.

In order to develop a collective whole from the individual contributions the authors were asked to comment on and relate to a preliminary paper presenting a foundational understanding of what could be meant by staging. Based on the comments received the piece was revised and appears as the present Chapter 2 in this volume.

The team of editors was set up to combine experience with the younger generation's fresh ideas and to combine internal and external perspectives. The editors have reviewed all of the chapters in the format of three-page abstracts for selection, and through two rounds of review of the full version of the chapters. This process has enabled a sound cross-breeding and mutual inspiration across the chapters and led to a well-integrated volume grounded in a variety of cases and approaches, but all contributing to the same project.

We hope this book will be of inspiration to the emerging groups of professionals facing sociotechnical challenges in the work with design and innovation and their educators.

Copenhagen and Lausanne, June 2020
Christian Clausen, Dominique Vinck, Signe Pedersen and Jens Dorland

PART I

Introducing staging

1. Staging collaborative design and innovation – an introduction

Jens Dorland and Dominique Vinck

Staging collaborative design and innovation proposes the concept of *staging* as a relevant tool and sensitizing device for the planning and facilitation of design and innovation activities by offering a repertoire of actionable collaborative strategies to intervene in and shape design and innovation processes. The book is a response to the increasingly complex challenges new types of professionals such as design engineers as well as urban planners or technology managers are facing designing products, services, systems or governing innovations and sustainable transitions. For instance, engineers and designers working to design or facilitate change in systems face a unique set of challenges in how to navigate competing interests and staging efforts, which have been ill-treated in existing research on co-design, staging, and other design approaches.

The staging concept take some of its roots in the theatrical metaphor in combination with inspiration from various academic traditions. In this first chapter, we will present and characterize the concept and shortly discuss its use throughout the different chapters in the book. We will see that it helps in avoiding a socio-cognitive reduction of design activity but opens new perspectives focusing the attention on the casting of the participants, the framing of the stage, and the resources it offers to them.

In Chapter 2, we will return to the origins of the staging perspective. The theoretical approach is based broadly within the STS (Science and Technology Studies) tradition that has studied the translation of sociotechnical networks to show how technology and innovation emerge and develop. However, besides literature on innovation and project management, little research has studied how these processes may be subject to intervention and active shaping. Innovation processes were studied by historians like Thomas Hughes (1983), social constructivists (Bijker et al. 1987), anthropologists (Downey 1998), and the actor–network theory (Latour 1987), but with no focus on design practices. A few STS or ethnographic studies investigated design activities describing them as situated processes, looking at the professionals and their negotiations to determine the form and function of new products (Bucciarelli 1996), pointing to the visual culture of engineering design (Henderson 1998), and the importance of drawing practices (Vinck 2003) and intermediary objects (Vinck 2012; Vinck and Jeantet 1995) and their equipping (Vinck 2011), but not the shaping of these situated activities. Thus, Signe Pedersen, Jens Dorland, and Christian Clausen explore more specifically the origins of the three structuring approaches that the staging approach builds upon: participatory design, shaping of technology, and organizational studies. The staging approach is a descendant of the Scandinavian tradition for action research from the 1960s and onward, the movement towards a proactive and action-oriented technology assessment in the 1980s (Hansen and Clausen 2002) and in particular the participatory design movement. One outcome is reflected in the following more actor-oriented spatial concepts intending to enable an emerging engineering design profession to stage strategic and political considerations and to include wider concerns in concept development. Here, the theatrical metaphor of staging became an inspiration helping to think about design as we will see in this book.

THE THEATRICAL METAPHOR

In theatre, the *stage* is the space where a performance generally takes place in front of an audience. The performance is expected to create emotion, reflection, awareness, learning, concernment, reaction, and maybe individual or societal change. The notion of stage focuses the attention to a few characteristics of the performance: its situated character; the arrangement of the scenery; the involvement of props,[1] costumes, decor and masking, flattage and traps, wing curtains and backdrop, set pieces; the dependence of the performance from the presence and the play of actors; the inclusion of the audience as passive or active participants to the performance, positioned into a specific placement, generally separated from the stage.

As far as designing a solution or developing an organization is considered as a kind of collective performance, it would be worthwhile to consider the design stage. This could help avoid reducing design activities to either a cognitive process (problem resolution, imagination and creativity, knowledge mobilization) or a socio-cognitive one involving various actors (mouthpiece of stakeholders, experts from various scientific or technical fields, potential users) acting according a specific method (a kind of script and the definition of some characters to which the involved people would identify themselves). The stage would also be important as a resource for the participants to define what is happening on an ongoing basis, locating the activity and framing its situated characteristics. Thus, grounding on the theatrical metaphor could open new perspectives on design and innovation, enriching our understanding of this activity with more items to consider than social and cognitive aspects.

As for theatre, the stage can be prepared. *Staging* means 'setting the scene' in a theatrical context, according to Pedersen, Dorland and Clausen (Chapter 2); below are some aspects worth pointing out regarding what staging generally is:

- **An overt and dynamic process** which cannot be reduced to a script, a plan, or guiding ideas (the text in theatrical context) framing the action to take place in a space spanning the performance.
- **Strategic action by one or various actors** resulting in an intended process where a *stage director* (either a creative expert, engineer, or other professional who has a structuring role), or a *humble* facilitator of collective explorations, enables the empowerment of actors and technicians, inviting and giving voice to not only the original text and to its own intuitions or ideas but also to his/her collaborators, stage managers, technicians, and actors.
- **Translation and interpretation** of a play (from a text), or some guiding ideas, into action, enabling and constraining the performance in time and space.
- **Composition or configuration, casting, and invitation** or enrolment in actor–network theory (ANT) terms (Callon 1986) of actors, definition of their role and associated characters (identities, actions, interactions).
- **Design of a 'scenic' space and its production**; this process encompasses objects or scenic elements like props, wings and set pieces, lights, and sounds that also play an active role in the staging, and their assembling (configuring and bounding them together in order to enable the performance of the actors on a scene).
- **Planning for rehearsals and performances**, which is the focus on the back-stage and in between plays.
- **Responses and navigation to unforeseen events**, change of players and set pieces, adjustments based on reviews and receptions between rehearsal and performance.

Pedersen, Dorland, and Clausen ask who and what is staging, and propose not to limit a priori the answer to the original text and the stage director, but to include also the different human actors and scenic elements. The theatrical metaphor of staging invites consideration of not only the composition, casting, and enrolment of actors and scenic elements, but also the design of socio-technical spaces (Clausen and Koch 2002; Clausen and Yoshinaka 2007) and their boundaries (i.e. who and what is invited to the stage and who and what are left outside). All these components, their composition, and assembling would influence the potential performance. In this sense, staging is concerned with how diverse actors, material objects, and ideas are drawn together in a common space to enable a performance.

Staging involves multiple actors. Traditionally, focus is put on the *stage director* (Pluta 2017) who designs the stage, with the help of a scenographer, and guides the translation of the script into a performance. It also involves a *stage manager*, who finds props for the production of the spectacle, organizes their realization, test, and reception, organizes their disponibility in space (on the stage or in the backdrop) and time (for use in the right moment), and their manipulation. This means planning the mobilization of technicians and operators and inspecting every detail prior to rehearsal and performance to ensure every part is ready to be used. During the performance, the stage manager must watch what is going on, anticipating potential incidents, and reacting to unexpected events. Finally, after the performance, the stage manager must ensure that everything is at its exact placement, in good condition and ready to be re-used, and maybe manage their repair and maintenance. Following these processes helps to understand that a performance is not so much the 'simple' execution of a written piece or the following of guidelines, but their local and situated application (Suchman 2007) and the resulting shaping and staging of encounters between multiple actors and scenic and non-human components. Thus, staging relates to how the performance is prepared and arranged; it refers to the actors who are invited onto the stage to frame creative ideas.

All these practices observed in theatrical experience are potentially useful to think about design as a collective, cognitive, and material activity. The different chapters in this book are dedicated to specific aspects of design and innovation exploring heuristical potentials of the theatrical metaphor. Various aspects we just mentioned here of the theatrical experience demonstrate their relevance for design and innovation.

In Chapter 15, Dominique Vinck and Mylène Tanferri will return to the metaphor of staging in order to point to its risks and limitations and then to enrich the notion thanks to both a field study on staging a performance and a literature survey on the theatrical experience. Learning from both field observation and the history of theatre, the chapter extends the perspectives on staging and draws up some lessons from the book. It reports on theatre's practices and evolutions highlighting its diversity and thus avoiding speaking about staging as a homogeneous reality and being trapped by a simplistic metaphorical reference.

And in the same way as Chapter 15 takes stock of the fruitfulness of the metaphor, Yutaka Yoshinaka and Christian Clausen in Chapter 16 offer a reading of the contributions across the book chapters from a shaping of design and innovation perspective, asking which staging actions and strategies can be identified across the book.

USE AND CONCEPTUALIZATION OF STAGING ACROSS THE BOOK CONTRIBUTIONS

The book presents a coherent collection of research papers where staging has been practiced, studied, and reflected upon in a variety of settings such as technology and product development, system transitions, organizational development, and the development of infrastructures and city life. The ideas published in this book are to a large extent based on a research environment engaged in the development of a new engineering curriculum informed by STS and engineering design principles at the Technical University of Denmark from 2000 and continued at Aalborg University from 2012 (Brønnum and Clausen 2013; Clausen and Gunn 2015; Clausen and Yoshinaka 2007; Dorland et al. 2019; Hansen and Clausen 2017; Jørgensen and Sorensen 2002; Pedersen 2020; Petersen and Buch 2016).

The collection offers theoretical and conceptual developments and empirical cases concerned with the staging of collaborative and cross-disciplinary design and innovation processes and together illustrates this *new actionable perspective on the staging of design and innovation.* The various chapters are thus relevant for engineers, urban planners, designers, product developers, and the increasing number of hybrids that combine the different professions. As it also pushes forward the study of design and innovation activities regarding their own shaping, it is also a contribution to academic research on design and innovation.

Thus, the book contributions address two aspects of staging. One is concrete guidelines and heuristics based on rich empirical cases that can potentially directly inform practitioners of ways to do things. The other aspect is conceptual developments resulting in concepts and ideas that can act as sensitizing devices for understanding different aspects of staging. Staging as an analytical notion invites pushing further the investigation on various unseen processes like the invitation to stakeholders or their equipping (see Chapter 14), and consequently nurtures further guidelines.

All the contributions are sustained by case studies. Some have the nature of action research where the authors actively exercise the staging approaches; some authors are practitioners from the field itself, while others study the staging efforts of others. Some cases are longitudinal, spanning months or years, while others take as their starting point one or a couple of delimited events. This diversity helps to explore different aspects of the staging in design and innovation. It leads to the identification of a variety of processes that need to be considered, explored, documented, and critically evaluated.

The book starts out in Chapter 2, which, as mentioned, explores the three structuring approaches that staging draws inspiration from, mapping a first collective understanding and conceptual frame, leading to a repertoire that can be drawn upon. The chapter sketches out a repertoire for action and insight composed of three aspects. First is the possible space or scene for staging that involves a process where a stage director designs or identifies a scene and invites actors, as well as the actual play that entails facilitation, navigation, reframing, and other activities in the performance of staging (Pedersen 2020). Second is the role of objects and materiality for interactions, also involving configuring the space to construct agency for different types of activities (Dorland et al. 2019), and enabling the travel of knowledge, narratives, discourses, and other aspects related to understanding and awareness that is essential for design and innovation. And, lastly, is the theatrical metaphor that can be seen as a sensitizing device that acts as a unifying imagery that allows for navigational considerations. The conceptual frame and outlined repertoire are flexible, and the book chapters use it in many different ways, and in the rest of this chapter we will present the different approaches to staging related to the core sections of the book:

- Part II: Staging participatory co-design with multiple actors
- Part III: Staging changes in networks and organizations through design of spaces and events
- Part IV: Staging interactions between research and innovation
- Part V: Staging experimentation and learning.

The case studies supporting the study of staging also relate to different sectors. In the book, they mainly concern **healthcare** (Pedersen and Brodersen in Chapter 3 on insulin conditioning and distribution, Sanders in Chapter 4 on service design and delivery, Pedersen and Brodersen in Chapter 5 on technology at a nursing home), **sustainable development** (Clausen and Gunn in Chapter 7 on indoor climate practices, Huulgaard et al. in Chapter 8 on transition toward circular economy, Dorland in Chapter 6 on renewable energy, Jensen in Chapter 12 on energy practices in households), **industry** (Brønnum and Clausen in Chapter 9 on product development, Bates and Juhl in Chapter 10 on public–private development projects and knowledge infrastructure, Bates in Chapter 11 on the transition from invention to commercialization), and **urban development** (Johansen and Lindegaard in Chapter 13 on urban nature, Hoffmann and Munthe-Kaas in Chapter 14 on urban planning), but, of course, the approach would be relevant for many other sectors in industry, services, and social innovation (Dorland in Chapter 6).

PART II: STAGING PARTICIPATORY CO-DESIGN WITH MULTIPLE ACTORS

This part contains three chapters applying participatory design within healthcare in relation to network building and product design in different ways. The healthcare context is characterized by being comprised of a complex array of multiple and diverse technologies and actors like hospitals, general practitioners, pharmacies, patients, and national health institutions. Furthermore, institutions play a central role and regulatory structures frame what is possible to change (Kensing et al. 2004).

The first chapter in this section, Pedersen and Brodersen (Chapter 3), illustrates staging through a case on the development of an insulin system in India by two design engineers, involving diverse actors from a company, representatives from the local healthcare sector, and diabetes patients in the Indian slum areas of New Delhi. This is done through a series of workshops focusing on developing the network and relations and not the physical products. The chapter contributes with a conceptual framework and staging guidelines for using it. Sanders (Chapter 4) illustrates how a facilitator can use a co-design approach to stage different levels of engagement in the front-end of the design process ranging from an intensive singular co-design event to a co-design culture with participatory relationships that grows over time. In contrast to Pedersen and Brodersen (Chapter 3), this chapter focuses on design of material artifacts and is based on decades of experience and not a specific case, representing a more hands-on and down-to-earth chapter contributing with specific guidelines. Lastly, Pedersen and Brodersen (Chapter 5) contributes with a case focusing on the design process of healthcare technology by a group of design engineering students at a nursing home that questions the complexity related to the involvement of multiple actors in collaborative design processes. The chapter employs an interesting and inspirational methodology that shows how students design and circulate objects for negotiation with elderly citizens with dementia and their caregivers. The chapter contributes to the conceptualization of front- and back-stage as part of the repertoire in design processes.

In the design setting that these chapters illustrate staging relates to how a co-design activity is prepared, framed, and arranged (Brodersen et al. 2008), and refers to how actors are invited onto the stage to frame problems, negotiate concerns, and enact circumstances using props such as design games, mock-ups and prototypes to form narratives (Brandt et al. 2005; Iversen et al. 2012; Pedersen 2020). Thus, casting the actors, inviting them, designing the props and having an idea of the object of negotiation is part of the staging, and the role of objects is essential for sharing knowledge and mutual learning.

This is especially well illustrated in Chapter 5, which contributes with the conceptualization of a framework expanding on the idea of the back- and front-stage that focuses on the development, inscription, and circulation of intermediary objects and provides insight into the iterative negotiations of both the matters of concern (including emerging ones) and solutions. The chapter also conceptualizes a new type of *fluid* intermediary objects that continuously evolves but remains back-stage. Examples include objects such as affinity diagrams and design specifications that have the potential to represent actors' concerns and thus at the same time also frame the solution space. They also inspire and frame front-stage activities while bridging the gap to the conceptualization of solutions. This framework enhances our understanding of the design process. This could be translated into specific guidelines like those of Sanders (Chapter 4) that are object-focused through a term called embodiments. The basic components of the guidelines are the actors, the place, the plan, and these embodiments that emerge in the execution of the staging plan by the actors in the place. Sanders uses these embodiments of participatory prototyping to engage the participants and facilitate the events, which are a special type of co-created objects that range from dreams, ideas, and concepts to full scale physical embodiments. Chapters 4 and 5 together thus provide a package of concepts to enhance understanding of the process and practical guidelines for planning and execution.

Pedersen and Brodersen (Chapter 3) provides both a conceptualization of a framework termed 'staging negotiation spaces' that entails an iterative process of negotiations and reframing in the creation of new networks, and guidelines for staging them. In contrast to the two other chapters the focus is thus more on the relational aspects. The chapter shows how the circulation of objects in as well as between spaces plays a central role in network building. The focus is therefore not the physical products that flow in the insulin system, but the creation and development of the relations between the involved actors.

The chapters thus contribute to staging in different ways. Chapter 3 sees the staging metaphor as encouraging the exploration of design and innovation as an iterative series of temporary negotiations with multiple actors, breaking with the perception of design as a linear process that can be planned, and with the users as passive audiences. Sanders in Chapter 4 sees staging as planning and facilitation of events, not drawing explicitly on the metaphor, but contributes with insight into different levels of engagement ranging from what can be achieved in a one-time workshop compared to a process spanning several events. Lastly, Chapter 5 takes the metaphor literally focusing on the front- and back-stage and what it implies, a clarification that is very relevant because many biases and unintended consequences in design processes, such as gender bias, originate from the design of the prototypes and the inherent scripts presented to users. The framework brings attention to the fact that prototypes designed back-stage can be intended to support the front-stage design process, and the focus this conceptual framework brings can help to challenge such biases.

PART III: STAGING CHANGES IN NETWORKS AND ORGANIZATIONS THROUGH DESIGN OF SPACES AND EVENTS

Staging has increasingly been a topic in relation to transition design as well as organizational and systemic change which often operates on a different scale than participatory design. This part focuses on the development and change of organizations and networks in relation to transition processes in society, and thus focuses on development over long periods of time. The difference from the previous section focusing on instances of development and design is an increase in the temporal scale and number of actors to focus on the process of organizing, of creating or changing organizations and society.

Dorland (Chapter 6) takes a marriage of organizational theory, STS, and participatory design and applies it to a study of how the founders of two local initiatives of international social innovation networks go from an idea to creating a network that they over time succeed in formalizing into organizations, as well as adapting and surviving despite set-backs. The chapter is a study of how staging is structuring organizational dynamics for maximum societal impact can be done, resulting both in practical guidelines in the form of strategies for organizational development, as well as conceptual developments of back- and front-stage focusing on the activities taking place between performances. Clausen and Gunn (Chapter 7) is based on the social shaping of technology perspective, where a major concern has been with the identification of situations and spaces where the shaping of technology could be analysed and political options and choices investigated (Clausen and Koch 2002; Russell and Williams 2002). The chapter takes as its starting point a case on engineering indoor climate in the building sector, studied through a series of workshops staged by researchers with engineers from different companies, with the intention of designing a transition, i.e. having an impact on the practices of engineering indoor climate in the industry. The chapter gives insights into the outcome of such workshops and contributes to the development of 'temporary spaces' as a concept for staging. Huulgaard et al. (Chapter 8) takes a similar approach building on this concept of temporary spaces in a case where a research project is trying to facilitate a transition to circular economy in an engineering company through the deliberate design of a contextual setting for experimentation. The contribution is mostly conceptual, pointing to the cumulative learnings and circulation between spaces that allows adapting the staging approach. The chapter illustrates the importance of taking advantage of emergent openings to create temporary spaces where it is possible for actors to experiment.

The focal point of the section is the 'spaces', referred to as temporary spaces, sociotechnical spaces, events, or scenes if drawing on the metaphor, and the activities taking place between them. These spaces or events are by Dorland (Chapter 6) seen as scenes where the performance that has been staged and rehearsed between the events plays out. The analysis focuses on events that over time form a structure through the objects the interactions in these spaces produce, like contracts, statutes, videos, reports, and action-plans etc. A crucial feature of this analytical perspective is the focus on actions and activities, in contrast to actors. In contrast to Huulgaard et al. (Chapter 8) and Clausen and Gunn (Chapter 7), many of the staging efforts identified by Dorland relate to the activities between spaces, like prepping allies, negotiating concerns, creating ideal conditions for the type of output aimed for, etc. Clausen and Gunn (Chapter 7) stage temporary spaces to enact knowledge by bringing in objects like narratives and field studies to challenge and reframe taken-for-granted understandings in the industrial field of indoor climate to make the production of specialized engineering knowledge more participatory. The staging efforts here focus on bringing attention to political processes in the creation of boundaries not only delimitating but also enabling certain innovation processes. The temporary spaces concept is also further developed as a focusing instrument and sensitizing device for studies of the staging of innovation to have a transition impact. And Huulgaard et al. (Chapter 8) lastly illustrates temporary spaces as an approach for enabling interaction and experimentation between stakeholders in a company setting. The case shows the importance of leveraging the opportunities that arise in an uncertain and ever-changing field of opportunities, helping to conceptualize a form of flexible and emergent staging that can inspire practitioners working with sustainability in companies.

The contributions in this part bring insight on political processes and how objects can be staged to help knowledge move across boundaries (Chapter 7), practical insight on navigation in a company setting by taking windows of opportunity in staging and enabling cumulative learning between spaces (Chapter 8), and a macro-level perspective with cases spanning several decades and a process view on organizations through adoption of the 'action-net' concept that provides insight on strategies and practical approaches when creating or developing organizations especially in relation to social innovation and social movements (Chapter 6).

Staging itself is understood very widely in this part. Dorland conceptualizes on the front- and back-stage as well as scenes in relation to understanding the ongoing staging of interactions in developing organizations. He further juxtaposes staging with that of sensemaking, where staging is the overt and strategic effort and sensemaking is an analytical concept employed to understand the interactions between actors. The staging efforts, the rehearsal, taking place behind the scene to a large extent are efforts to have an impact on the sensemaking process happening during a performance. Clausen and Gunn as well as Huulgaard et al. also understand staging as an overt activity, but use the metaphor itself more lightly, seeing it as mostly a sensitizing device for use in planning and preparing 'temporary spaces'.

PART IV: STAGING INTERACTIONS BETWEEN RESEARCH AND INNOVATION

The concept of staging has also gained attention in engineering design. For instance Andreasen et al. (2015) dedicate an entire chapter of their book *Conceptual Design* to staging where they not only assign staging to be an important skill of the individual designer but also to the design team. This part of the book has chapters focusing on product and technology developments in various engineering companies with object and practice focused perspectives. The three chapters work with objects and materiality in staging as well as the political processes in mature engineering organizations, contributing with strong empirical illustrations of how staging unfolds. The chapters also conceptualize various terms like development constitution and referential alignment that help us to understand the sociomaterial configuration inherent to this type of organization.

The first chapter by Brønnum and Clausen (Chapter 9) studies how the Front End of Innovation (FEI) and development activities aimed at developing new ideas and product concepts 'in front of' the product development activities are enabled by project managers, based on an in-depth study of a development project in a mature engineering company. The chapter conceptualizes FEI as a development space and focuses on staging as a political process aimed at enabling explorative and more radical innovation. Bates and Juhl (Chapter 10) focus on the 'what, how and why' of staging in a case of a public–private technology development project involving technology tests and the associated industrial–academic hybrid production of knowledge. The focal point is on the project manager who is also one of the authors of the chapter, providing practitioner insight, and the object of staging is a test setup with an intention to create a knowledge infrastructure that can align the interest of the involved partners. Bates (Chapter 11) continues the focus on objects in engineering design and development, contributing with a strong auto-ethnographic empirical case study on the active use of 'intermediary objects' in the practices of managers tasked with staging transitions from invention to commercialization. The chapter focuses on the creation and staging of these objects, which might be a whitepaper or business case, contributing with practical insight for engineers and managers involved in commercialization of innovation.

The two main aspects in this section are the materiality and objects in engineering practice, and the political processes of development organizations. Clausen and Brønnum focus mostly on the political processes through the further development of a concept called 'constitution of development'. The chapter contributes with four different strategies for how aspects of such a constitution can be enacted in various ways to create different types of development spaces, which is the object of staging in their case study. Bates and Juhl focus on the material aspect with a case on a test setup for a new non-petroleum lubricant, which was a key object critical in the early connection of partners around 'common performance criteria'. Through the process of establishing the involved interests and configuring a new process of knowledge production, the test setup represents a type of knowledge infrastructure that the chapter conceptualizes as 'referential infrastructure'. This concept focuses on how to enable knowledge artifacts and processes to refer between different material, organizational, local, and temporal settings. It is a complex conceptual development that also operates with several distinct object types that gives insight into how project managers can stage with material artifacts. Bates (Chapter 11) in the last chapter in this part focuses on both the political processes and materiality in engineering in a case study on a technology development project following the messy translation processes taking place when going from invention to commercialization. The chapter focuses on how objects can be used for staging within engineering management practice, focusing on the practical experience in project management and the insight that can be drawn from it. This analysis leads to several practical insights for engineers that can enhance their means of actions and navigate the complex process of negotiations with other actors. Staging as understood in these three chapters is seen as both the strategic planning and preparation, the performance, and the outcome, with a strong focus on both material objects and political processes. None of the chapters draws strongly on the metaphor itself, but especially the two first chapters contribute with new conceptual developments enhancing the action-oriented approach to staging, while the last chapter finishes with a case illustrating several of these insights in practice.

PART V: STAGING EXPERIMENTATION AND LEARNING

The penultimate part of the book focuses on staging urban living and planning through interventions and experimentation like Living Labs, workshops, and other types of spaces, and is thus relevant for both governance and transition research. The chapters work with very different temporal perspectives, from Johansen and Lindegaard (Chapter 13), who conduct two workshops within months of each other, to Jensen (Chapter 12), which works intensely with energy practices in households over several months, to Hoffmann and Munthe-Kaas (Chapter 14), which draws on a 50+ year perspective in urban planning.

Jensen (Chapter 12) shows how researchers can stage interventions in resource-intensive practices through Living Labs. The chapter discusses the different steps involved in staging Living Labs in Denmark, and what a practice-theoretical understanding contributes to the concept of staging. Staging is seen as the strategic planning of these Living Labs with a focus on social and material practices. Johansen and Lindegaard (Chapter 13) look at interventions in relation to the bureaucratic practice in the realization of a strategy for urban nature in the City of Copenhagen. The operational definition of urban nature changed significantly between different locations in the organization, which was seen as a challenge. This resulted in an interest in exploring the concept of urban nature, leading to a series of workshops planned and facilitated by the researchers exploring existing ideas and practices while simultaneously challenging existing problematizations. Hoffmann and Munthe-Kaas (Chapter 14) continue the focus on urban planning based on three cases on facilitating experimental spaces staged by the authors, arguing that the field is dominated by hegemonic modernist ideas that frame how lives can be lived. They argue that opening the planning processes up through staging processes that attempt to make the ordinary controversial can point towards alternative ways of living in cities in the future. Using the concepts of 'ontological choreography' and 'agonistic pluralism', they especially emphasize the need to carefully craft the 'invitation' to the stage, which they explore through different devices, and to inspire and challenge dominant perceptions of urban life and futures. The focus of this part is strongly on staging experimentations and facilitating learning through spaces that aim to have a specific impact that changes current practices and conceptions of specific actors.

This contrasts with the spaces in Part III that focus on developing objects that over time create new networks or change old ones to have an impact on the development of organizations and society. The three chapters in this section, however, approach this challenge very differently. The intervention staged by Jensen was initiated and guided significantly by the researchers, which is a contrast to participatory design approaches in Part II. The practice-theoretical perspective was also heavily engaged in framing the spaces as well as configuring connections between those who were invited onto the stage. The practice-theoretical perspective offers *different* explanations and thus different related *spaces for change* in contrast to behaviour change theories that rely mainly on rational consumer choice models. In contrast to the other staging approaches represented so far, the practice-theoretical perspective focuses on the continuous activities of daily life, here the practice of washing clothes among others, and not on a one-time performance like a specific product development project. The focus is thus moved away from the planning and rehearsals of a play to the ongoing performance of everyday life. The Living Labs are staged to bring the researcher closer to this everyday life, where many other design approaches focus on short-term interaction in the design phase. Johansen and Lindegaard illustrate a more delimited approach with two workshops that try to create insights for the participants on their own practices by challenging and exploring conceptions of urban nature inside the organization. The two workshops lead to a re-evaluation of organizational actor positions in urban nature planning and reveal how researchers, the subject matter, and other actors could be shifted to new positions, enabling new agencies. The chapter illustrates a very practical approach where physical objects are brought into the workshops to represent nature in different ways. The chapter also challenges the metaphorical notion of staging by introducing the notion of environmental theatre as a guide for the preparation, execution, and interpretation of these workshops. Lastly, Hoffmann and Munthe-Kaas represent a conceptual approach focusing on the aspect of the invitation, based on the development of the 'Thinging' concept emphasizing how urban places and processes can be understood as spaces of controversy and democratic exchange rather than problems that have correct (engineered) solutions. By seeing urban design as a Thinging practice Hoffmann and Munthe-Kaas find that new spaces can be opened for the political to be (re)introduced in urban planning. They explore how staging can be developed as an approach to create Things that can engage a multiplicity of actors in the development of the cities of the future. The chapter thus contributes with an analytical frame developed to focus on how the crafting of invitation matters for the stakeholderness: who is invited on the stage, and what futures can be explored? The chapter provides insight on how different invitational devices configure participation differently.

PART VI: REFLECTIONS – HOW STAGING IS UNDERSTOOD AND USED

Across the different chapters, the focus on staging has led to different observations and identifications of dimensions and processes, according to specific dynamics of the cases under study and to the objectives and ambitions in the cases. Furthermore, the ways to think about staging range from a metaphor used as a sensitizing device but otherwise only loosely coupled to both action-oriented and analytic frameworks building heavily on the theatrical vocabulary. The lingering question is what is staging, what may it perform, and what motivates its application by practitioners?

The converging contribution of all the chapters is that they show staging as something overt and strategic, often relating to casting and involvement of actors, designing and preparing events and the scenic space, but also to the creation of objects and scenic elements as a form of props that form the conditions, actor equipping, and infrastructure for an envisioned development. Staging broadly refers to 'setting the stage', but, as pointed out by Vinck and Tanferri (Chapter 15) the rehearsal is important not only as facilitation during the play and the final performance, i.e. when a product enters daily practice for instance, but also for the preparation of design activities themselves. Not only spaces, events, and props but also the involved actors in design activities need to be equipped and prepared through specific rehearsal of design activities. If one chapter focuses on one aspect or another, this often relates to the temporal and spatial scale of the cases. As Dorland (Chapter 6) focuses on creation and development of organizations over decades, the rehearsals and back-stage planning take the high-seat, while the facilitation during the specific events where the play unfolds are largely black-boxed. Jensen (Chapter 12) likewise illustrates the focus on rehearsal and stage-setting by preparing props and actors to steer the ongoing play of everyday energy practices.

What staging performs, broadly, is sensitivity to how agency is constructed helping practitioners in shaping of design and innovation towards desired outcomes considering multiple and diverging concerns. Staging is thus an action-oriented approach to utilizing the sensitivity to agency ANT brings (see Chapter 2), and the motivation for practitioners is this awareness and sensitivity that make agency an object for staging. From reading across the chapters of this book Yoshinaka and Clausen (Chapter 16) point out that staging is applied to a variety of problems and issues as well as the prospects for solving them.

And this final chapter (Chapter 16) reflects on the contributions by drawing up and elaborating on concrete strategies for staging in relation to the different types of activities covered in the chapters. These activities range across creating workshops, facilitating co-design processes, impacting urban planning, changing practices, project management, developing organizations, navigating inside institutions, etc. The findings range from very practical guidelines as exemplified by Sanders (Chapter 4), to more middle-ground strategic recommendations like Dorland (Chapter 6) or development of frameworks in Pedersen and Brodersen (Chapter 3), to more conceptual developments and considerations like the idea of Thinging (Chapter 14) and referential alignment (Chapter 10) that can act as sensitizing devices, enriching the staging vocabulary in relation to specific fields and activities. Lastly, several chapters also illustrate that theatre is not just theatre, it can be worthwhile to consider what type of theatre is being rehearsed and played, like Johansen and Lindegaard (Chapter 13) that builds on the idea of environmental theatre, Pedersen and Brodersen (Chapter 3) that created a framework over improvisational theatre, while Vinck and Tanferri (Chapter 15) points to the diversity of staging practices and challenges we can learn from the theatrical practices, e.g. regarding the importance of rehearsal, the involvement of the spectators, and the staging of light and sound. The metaphor is thus rich and flexible but should not be followed slavishly but rather enable reflection. The parts of the conceptual framing and repertoire outlined in Chapter 2 of relevance depend on the context and situation, and the fruitfulness and dangers of the metaphor is outlined in Chapter 15. Chapter 16 reflects on the strategies across the book from a shaping perspective, from which the strategies can be seen as concerned with four different aspects: (1) staging applied to strategically bridge established networks to seek transitioning systems of knowledge or new innovative possibilities; (2) staging concerned with the political navigation and enactment of commitment of agency to emerging spaces for front-end innovation and transitions; (3) deliberate development and circulation of objects through transformative spaces that can prompt articulation and representation of concerns while dealing with shifting alignments of objects as well as actors; and (4) strategies of staging as experimentation demonstrating how preconceived ideas can be challenged. Above all, Yoshinaka and Clausen (Chapter 16) emphasize the role of political navigation in the wider scenery or arena where staging takes place among the repertoires of staging offered across the book, and see developing the navigational skills of the professionals in the ever-changing condition of development as crucial.

NOTE

1. Furnishings and other large or small items which are not part of the scenery, among which are hand props or personal props which are kept in a player's costume.

REFERENCES

Andreasen, M.M., C.T. Hansen and P. Cash (2015), 'Staging conceptualization', in *Conceptual Design*, Heidelberg: Springer, p. 394.

Bijker, W., T. Hughes and T. Pinch (1987), *The Social Construction of Technological Systems: New Directions in the Sociology and History of Technology*, Cambridge: MIT Press.

Brandt, E., M. Johansson and J. Messeter (2005), 'The design lab: Re-thinking what to design and how to design', in T. Binder and M. Hellström (eds), *Design Spaces*, Helsinki: Edita Publishing Ltd, pp. 34–43.

Brodersen, C., C. Dindler and O.S. Iversen (2008), 'Staging imaginative places for participatory prototyping', *CoDesign*, **4** (1), 19–30.

Brønnum, L. and C. Clausen (2013), 'Configuring the development space for conceptualization', in *Proceedings of the 19th International Conference on Engineering Design (ICED13)*, vol. 3, Seoul: Design Society, pp. 171–80.

Bucciarelli, L. (1996), *Designing Engineers*, Cambridge, MA: MIT Press.

Callon, M. (1986), 'Some elements of a sociology of translation: Domestication of the scallops and the fishermen of St Brieuc Bay', in *Power, Action and Belief: A New Sociology of Knowledge?*, London: Routledge, pp. 196–223.

Clausen, C. and W. Gunn (2015), 'From the social shaping of technology to the staging of temporary spaces of innovation – A case of participatory innovation', *Science and Technology Studies*, **28** (1), 73–94.
Clausen, C. and C. Koch (2002), 'Spaces and occasions in the social shaping of information technologies: The transformation of IT-systems for manufacturing in a Danish context', in K.H. Sørensen and R. Williams (eds), *Shaping Technology Guiding Policy: Concepts, Spaces and Tool*, Cheltenham, UK and Northampton, MA, USA: Edward Elgar Publishing, pp. 215–40.
Clausen, C. and Y. Yoshinaka (2007), 'Staging socio-technical spaces: Translating across boundaries in design', *Journal of Design Research*, **6** (1–2), 61–78.
Dorland, J., C. Clausen and M.S. Jørgensen (2019), 'Space configurations for empowering university-community interactions', *Science and Public Policy*, **17** (5), 689–701.
Downey, G. (1998), *The Machine in Me: An Anthropologist Sits among Computer Engineers*, New York: Routledge.
Hansen, A.G. and C. Clausen (2002), 'The role of TA in the social shaping of technology', in G. Banse, A. Grunwald, and M. Rader (eds), *Innovations for an E-Society*, Berlin: Edition Sigma, pp. 91–8.
Hansen, P.R. and C. Clausen (2017), 'Management concepts and the navigation of interessement devices: The key role of interessement devices in the creation of agency and the enablement of organizational change', *Journal of Change Management*, **46** (5), 344–66.
Henderson, K. (1998), *On Line and On Paper: Visual Representations, Visual Culture, and Computer Graphics in Design Engineering*, Cambridge, MA: MIT Press.
Hughes, T.P. (1983), *Networks of Power: Electrification in Western Society, 1880–1930*, Baltimore: Johns Hopkins University Press.
Iversen, O.S., K. Halskov and T.W. Leong (2012), 'Values-led participatory design', *CoDesign*, **8** (2–3), 87–103.
Jørgensen, U. and O.H. Sorensen (2002), 'Arenas of development: a space populated by actor worlds, artefacts and surprises', in K.H. Sørensen and R. Williams (eds), *Shaping Technology, Guiding Policy: Concepts, Spaces and Tools*, Cheltenham, UK and Northampton, MA, USA: Edward Elgar Publishing, pp. 197–222.
Kensing, F., D.L. Strand, J. Bansler and E. Havn (2004), 'Empowering patients: PD in the healthcare field', *Proceedings of the Participatory Design Conference, Toronto, Canada, July 27–31, PDC-04*, **2**, 72–5.
Latour, B. (1987), *Science in Action: How to Follow Scientists and Engineers through Society*, Cambridge, MA: Harvard University Press.
Pedersen, S. (2020), 'Staging negotiation spaces: A co-design framework', *Design Studies*, **68**, 58–81.
Petersen, R.P. and A. Buch (2016), 'Making room in engineering design practices', *Engineering Studies*, **8** (2), 93–115.
Pluta, I. (2017), *Metteur En Scène Aujourd'hui – Identité Artistique En Question?*, Rennes: PUR.
Russell, S. and R. Williams (2002), 'Social shaping of technology: Frameworks, findings and implications for policy', in K.H. Sørensen and R. Williams (eds), *Shaping Technology, Guiding Policy: Concepts, Spaces and Tools*, Cheltenham, UK and Northampton, MA, USA: Edward Elgar Publishing, pp. 37–131.
Suchman, L.A. (2007), *Human–Machine Reconfigurations: Plans and Situated Actions*, Cambridge, MA: Cambridge University Press.
Vinck, D. (2003), *Everyday Engineering: Ethnography of Design and Innovation*, Cambridge, MA: MIT Press.

Vinck, D. (2011), 'Taking intermediary objects and equipping work into account in the study of engineering practices', *Engineering Studies*, **3** (1), 25–44.
Vinck, D. (2012), *Accessing Material Culture by Following Intermediary Objects*, London: INTECH Open Access Publisher.
Vinck, D. and A. Jeantet (1995), 'Mediating and commissioning objects in the sociotechnical process of product design: A conceptual approach', in D. Maclean, P. Saviotti, and D. Vinck (eds), *Designs, Networks and Strategies*, vol. 2, COST A3 Social Sciences, Bruxelles: EC Directorate General Science R&D, pp. 111–29.

2. Staging: from theory to action

Signe Pedersen, Jens Dorland and Christian Clausen

INTRODUCTION

The staging perspective on design and innovation in this book offers an action-oriented and participatory viewpoint on the processual unfolding of change, by providing analytical perspectives and concepts as well as actionable guidance and strategies for professionals such as design engineers, city planners, or technology managers to deal with a diversity of concerns while solving societal and so-called wicked problems that are characterized by being complex and contradictory. While other concepts like Design Thinking focus on tools and methods for changing the status quo, the novelty of the staging perspective is on combining an action-oriented approach from the Scandinavian tradition of participatory design and innovation with insights from Science and Technology Studies (STS) offering a strategic view on change.

The staging perspective presented in this book offers insights into the design of products, services, and larger systems, as well as infrastructures, urban development, and societal change and transitions. Thus, staging applies to different sizes of networks and promotes a perspective that ties together individual design projects with larger societal transitions, giving insights into the location of agency and connecting the activities at the local level with institutional changes. For instance, scaling up activities can be done by connecting local design projects with wider challenges of sociotechnical change and transitions.

By drawing on dramaturgical metaphors the space(s) where design and innovation take place is here compared with a stage where plays are performed. Traditionally in the world of theatre, the Scandinavian word for staging, *iscenesættelse* (meaning "setting the scene"), refers to the process of translating and interpreting a play from text or dramaturgical proposal into action in a scenic space where a performance takes place (Kjølner 2007). In contrast to Goffman (1959), who focuses on the stage (Belliger and Krieger 2016) in his use of the theatrical metaphor, this book focuses on *staging* as a central process in design and innovation. A key element in the staging process is the interpretation of an original material like a design challenge that evokes a certain framing selected for the play in question (here project or situation), which guides action and thus enables (and constrains) performances. See Vinck and Tanferri (Chapter 15, this volume) for a discussion of the theatre metaphor and its implication for staging design and innovation.

This chapter intends to illustrate the different aspects of staging and the research traditions it is based on, outlining an approach and providing a repertoire of ways to configure and facilitate interactions across actors and objects that is illustrated in different ways throughout this book (see Figure 2.1). In the following sections, we will draw up how we understand the staging of design and innovation, the vocabulary, the process, through a discussion with the literature on which we build our understanding. We will focus on key concepts and understandings of staging to inform a repertoire of approaches of how to make staging actionable for diverse professions.

THE FOUNDATIONS OF STAGING

Staging in Participatory Design

Participatory design originates from the democratically oriented Scandinavian design tradition advocating that direct involvement of people in co-design should take centre stage. The first participatory design projects focused on 'democracy at work' referring to researchers' and designers' engagement with, for instance, the labour unions in the 1960s and 1970s as a response to the introduction of information technology (IT) equipment such as computers in the workplace (Bjerknes et al. 1987). This engagement was concerned with balancing the resource deficits between management and workers and the core idea that people influenced by new technology should be involved in designing it (Ehn 2008; Simonsen and Robertson 2012). It still is.

Later participatory design researchers kept democracy as a central element in the design process by giving voice to people being marginalized in sociotechnical change (Björgvinsson et al. 2012) and, in the beginning of the new century, focus shifted towards the involvement of not only marginalized end-users but of multiple actors affected by a new design or situation: "Ideally, users at many levels participate so that change can be shaped from several perspectives" (Bratteteig and Gregory 2001). The central figure in the efforts of involving and engaging the diverse actors is the participatory designer, who actively organizes and facilitates design events such as creative co-design workshops (Binder et al. 1998; Brandt 2001).

Since the 1990s researchers from the participatory design community have been inspired by dramaturgical and theatrical metaphors to describe and

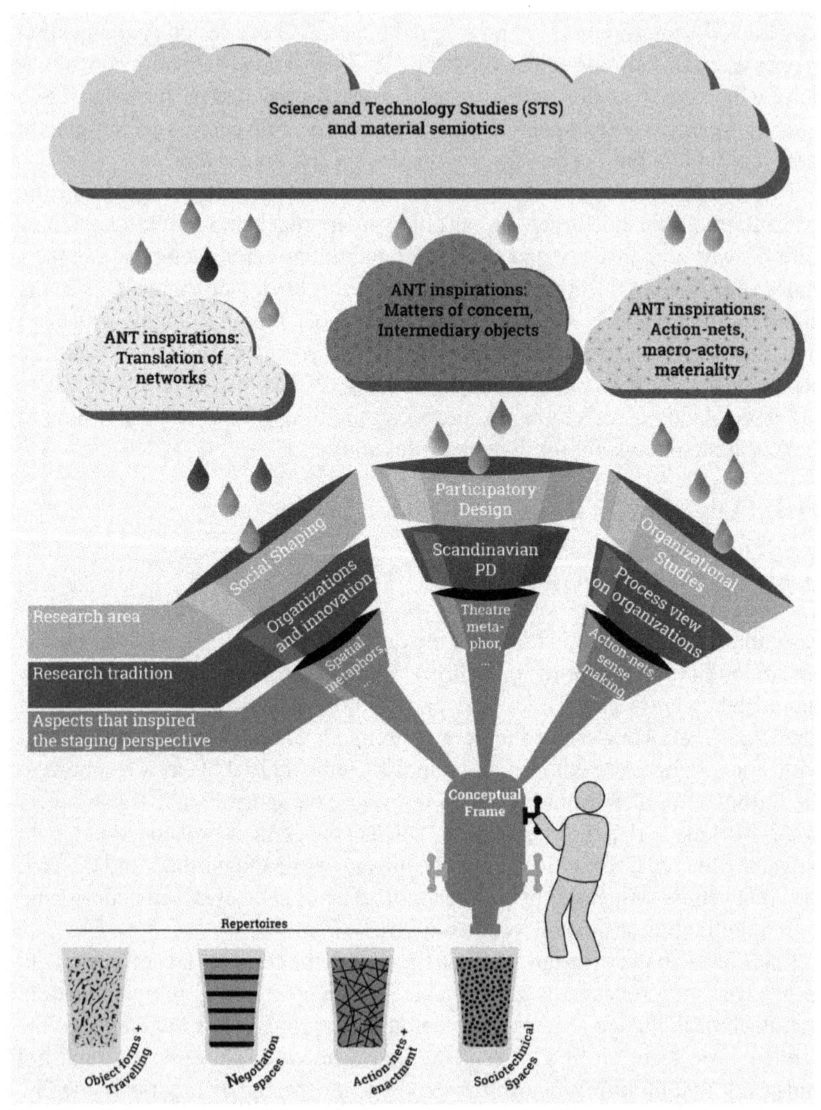

Note: Repertoires for staging design and innovation processes building on insights from STS and informed by actor–network theory (ANT) and further inspired by work within participatory design, shaping of technology and organizational studies.

Figure 2.1 Repertoires for staging and their origin

facilitate the interplay between designers and other actors that happens during events as a means of drama and props (Brandt et al. 2008; Ehn and Sjögren 1991). For instance Brandt, Johansson and Messeter (Brandt et al. 2005) experiment with engaging users in rehearsals of scenarios for possible imminent use through a process of staging, evoking and enacting future scenarios with props such as low fidelity mock-ups.

Typically staging refers to the process where designers carefully arrange and frame the collaborative event or activity (Brandt et al. 2005; Brodersen et al. 2008) by inviting actors to the stage and preparing props that also play a central role in the performance. Thus, part of this process entails selecting and inviting the actors and developing or selecting creative collaborative props and methods – for instance prototyping or design games – to facilitate actors' engagement as active participants in a process of mutual learning (Bødker et al. 2017; Bratteteig and Wagner 2012).

As methods are seen as central for participatory design, much attention has been devoted in the literature (as in e.g. Design Thinking) to methods and tools for engagement and design like in Micheli et al. (2018). The early Design Thinking literature investigated principles and practices of the professional expert designer including cognition and practical design methods, while the managerial strand of Design Thinking primarily promotes a toolbox of methods to be used in empathic problem identification, idea generation, and user testing (Pedersen 2020). But, while methods, approaches and tools, which are promoted in Design Thinking as well as in participatory design, are extremely valuable for designers, the tendency to focus on the methods in themselves only to a lesser degree caters for an informed actionable approach to participation that takes into account the wider perspectives "illuminat[ing] the context into which the designer is intervening" (Kimbell 2011: 287).

Thus, in recent years, members of the participatory design community have responded to this challenge by seeking inspiration within STS and particularly ANT. Through the lens of ANT, design is seen as a social and political activity playing a vital role in the shaping of our societies (Storni et al. 2015). According to Storni et. al. (2012) the contribution of ANT to participatory design is twofold as it offers both (1) an analytical perspective to describe and understand the sociotechnical settings a design project is immersed in, and (2) analytical tools to investigate and describe the design processes itself. To characterize the sociotechnical setting of a so-called Design Lab participatory design researchers such as Björgvinsson et al. (2010) and Stuedahl and Smørdal (2015) have used concepts from ANT to e.g. understand their research environments as *Things* (sociomaterial "collectives of humans and non-humans" through whom "matters of concern" or controversies are handled). And with her framework "Staging negotiation spaces" Pedersen (2020) draws upon the theatrical metaphor as well as on ANT to "emphasise the situated and political nature of negotiations, material and symbolic factors, and the role of interaction" and to present a vocabulary using the concept of intermediary objects (Boujut and Blanco 2003; Vinck 2012) to describe the role of props (e.g., methods, drawings, prototypes) once circulated to mediate in collaborative negotiations. In doing so, Pedersen stands on the shoulders of Bødker et al. (2017) who challenge the focus on single events such as workshops and conferences that they term frontstage activities. Drawing upon the work of Goffman (1959) and symbolic interactionism they use the concept of frontstage and backstage to draw attention to what they term the backstage work of design, which involves meetings, development of different materiality for later use on the frontstage, etc. In line with these arguments Pedersen's (2020) negotiation spaces framework encompass frontstage as well as backstage activities highlighting how staging in many cases is a collaborative effort rather than the work of a single creative designer.

Staging in the Shaping of Technology

Where participatory design focuses on the designer as key actor in the staging of participatory interaction, typically situated in a design project, the social shaping of technology (SST) (Sørensen and Williams 2002) offers a more analytical perspective on staging sociotechnical design and innovation seen as distributed activities typically involving entrepreneurs, innovation and project managers, engineers, designers, planners, policy players, etc. and a diversity of knowledge domains and interests. SST can be seen as an overarching term drawing on broad streams of STS including the Social Construction of Technology (Bijker et al. 1987) and ANT (Akrich et al. 2002; Callon 1986; Latour 1987).

A major concern within the SST has been with the identification of situations and spaces where the shaping of technology, innovation and design could be analysed and political options and choices investigated (Russell and Williams 2002; Sørensen and Williams 2002). Several concepts have been advanced aiming at characterizing and analysing such spaces for social shaping like strategic niche management and sociotechnical regimes (Geels and Schot 2007; Rip and Schot 2002) and sociotechnical constituencies (Molina 1997). These concepts have an ambition to characterize situations where choices are or can be made concerning the direction or outcome of technological development. Typically, SST scholars characterize and explore situations where technological networks or diverse worlds or domains of knowledge or practice interfere or interact, such as for instance in the crossing of boundaries between development and use (Clausen and Koch 2002).

However, while these concepts are mainly analytical and concerned with how design and innovation are generally being created and shaped, the staging approach presented in this book has been looking for a more actionable approach (Clausen and Gunn 2015; Dorland et al. 2019). This concern within the SST has in Scandinavia historically been developed through increasing instances of intersection and exchange with the participatory design movement helped by research and development programmes within action-oriented technology assessment (Hansen and Clausen 2002), participatory innovation (Buur and Matthews 2008), user-oriented innovation (Jørgensen et al. 2009) and the development of engineering curriculums within design and innovation (Andreasen et al. 2015). Here, spatial concepts have become oriented towards how design and innovation processes can be actively shaped to meet certain outcomes, to enable emerging engineering design and similar designerly oriented professions to stage strategic and political encounters, and to navigate concept development in arenas populated with actors, actor worlds, and surprises (Jørgensen and Sørensen 2002).

Clausen and Yoshinaka (2005, 2007) draw upon the SST tradition and examine the configuration of sociotechnical spaces where actors and material objects interact in sociotechnical and discursive processes of translation. They point at how staging is concerned with selective invitations to participation, the ongoing negotiation of boundaries, enacting underlying political discourses of participation and change which allow or encourage particular concerns to be voiced while others are excluded, thus offering a broader perspective to staging including a wider sociotechnical perspective than the participatory design tradition typically does. By drawing on political process theory concerned with processual change (Dawson et al. 2000) and ANT this strand of research has more broadly attended to the political navigation of organizational situations with an interest in how to include concerns outside the dominant perspectives in design and innovation such as working and external environment (Kamp 2000) etc. The organizational situations of concern would further typically include how actors such as project managers or engineering designers stage and navigate processes of change or the development of product concepts in the front end of innovation (Gish and Clausen 2013; Jensen et al. 2018), processes in which social players interact with one another from a diversity of departures, and with technological artefacts and management programmes for innovation and change. For instance, Clausen and Yoshinaka (2007) point at how particular artefacts and knowledge objects serve as boundary objects (Star and Griesemer 1989) which selectively translate knowledge from a variety of knowledge domains such as user experiences, expertise on specific materials or particular business logics into an engineering design or development project. The selection of such objects is key to the staging of boundaries to what they allow to pass and include and in which way they frame particular knowledge.

Metaphors typically used in SST are not so much the theatrical metaphor, but rather for instance sociotechnical spaces that may perform as sensitizing devices pointing at spaces such as a development project or the front end of innovation where technology and innovation could be analysed, inquired into, problematized and influenced by actors and perspectives other than the prevailing and dominant ones. They can be seen as attempts to turn the insights from STS and in particular ANT into an actionable perspective. By considering the inclusion or exclusion of actors, material and symbolic objects and concerns in a space and the construction of boundaries defining the space, staging attends to creating the possibilities for network translation and formation to happen.

Staging in Organizations

Drawing on ANT, a line of organizational studies characterized by a process understanding sees organizations as iteratively developing but only temporarily stable networks. This perspective is about democracy and empowering actors, as put by Czarniawska and Hernes (2005: 12), "ANT can empower micro actors, explaining to them the mechanisms of power creation". Part of the focus is thus on empowerment through new insight and actionable approaches.

Large networks that are perceived as actors are often termed macro-actors (Czarniawska and Hernes 2005). As explained by Callon and Latour (2015: 293), originally published in 1981, constructing a macro-actor and its power in relation to other actors is about "associating the largest number of irreversibly linked elements". A macro-actor grows in influence by building on a foundation of black-boxes as commented by Callon and Latour (2015 [1981]) "a macro-actor can be understood as a micro-actor 'seated on top of many (leaky) black boxes'", such as global warming, a management concept such as Lean, democracy or an organization like a university, company, or even the United Nations. These black-boxes are all actor-networks composed of heterogeneous elements like people, objects, discourses, regulations, technology etc.

ANT also contribute with a focus on materiality and objects illustrating that such black-boxes need to be anchored materially to take part in interactions (Dorland 2018; Law 2002; Law and Hetherington 2000), or, as put by Michael (2017), what can enter into an association are only elements in their concreteness and specificity. Macro-actors is an interesting concept because it explains how organizations can grow and gain agency over time, like the sustainable energy community on the island of Samsoe that grew over two decades and other "old" social innovation networks (Dorland 2018). It then also potentially makes macro-actors objects of staging activities.

What could be termed Scandinavian neo-institutionalism takes this perspective in a marriage with organizational theory building on symbolic interactionists like Weick (Czarniawska and Hernes 2005), which among others builds on Goffman, both drawing on the theatrical metaphor (Czarniawska 2006). This perspective conceptualizes the process nature of organizations as emerging from nets composed of associated actions building on Weick's sensemaking perspective (Belliger and Krieger 2016). Ontologically an organization is thus brought into being as it is performed (Robichaud and Cooren 2013), or as commented by Hernes (2008) actor-networks are forever in the making.

What is studied is thus not organization but organizing activities through events, which is then the object of staging. The events are the smallest observable temporal happenings (Allport 1954). ANT contributes with how events are connected through time and space by materiality (Czarniawska-Joerges and Sevón 2005; Strum and Latour 1987), while Weick focused only on linguistics. This is significant, as many of the phenomena focused on in organizational studies are notoriously hard to capture empirically (Pel et al. 2017).

Returning to the notion of staging, the central actors are the events of the past and how they constitute or can be configured into a structure (Dorland et al. 2019). The conception of action is thus focusing less on actors making things happen, the heroic protagonist, but more ecological seeing agency as dispersed (Gherardi and Nicolini 2005). Organizations involve hundreds or thousands of people that come and go, and it is the objects from events that they leave behind that acts.

The actors doing staging are then people with strategic intentionality in shaping events to reach specific outcomes, to configure the resulting action-net of an organization in certain ways. Many staging efforts thus lie between events, in planning, rehearsing, constructing, and other efforts to stage these events. In contrast, the traditions of participatory design and SST drawn upon here focus mostly on singular events or projects, and not the long-term development of organizations. Czarniawska also works on an institutional level, in contrast to the SST tradition here that often focuses on companies, giving insight into staging in a different context

Staging also takes place during events to facilitate, navigate, and shape outcomes, as staging is seen as all strategic and overt efforts no matter when they take place. This perspective, however, takes a more macro-view, seldom studying the internal details of singular events, and emphasizing the backstage preparations. Events can here be many types of spaces, from workshops to conferences, meetings, teambuilding, and any other interaction between multiple actors, which here and in participatory design are seen as the frontstage. This perspective is necessarily looking at longer temporal storylines, empowering actors in being strategic in the staging of events, to affect the ever-translating organization.

Common Grounds and Differences

At the root of the three strands that make up the origin of our staging perspective lies experience from Scandinavia, especially reflected in the participatory design tradition, but also visible both in the particular traditions in shaping technology and organizational studies drawn on here. These traditions all draw together perspectives on change, the occupation with how to handle, in more or less democratic ways, a variety of concerns and perspectives in a sociotechnical world.

Spaces seen as situations where problems and concerns can be negotiated, addressed, scrutinized, and dealt with form another common ground for the three traditions, but although there are clear differences in the way such spaces are conceptualized, from the negotiations in collaborative design projects, an arena for political contestation in an organizational setting, or as far-reaching processes of organizing.

The understanding of actors and agency is another contrast. Participatory design revolves around how designers stage their interacting with a multiplicity of relevant stakeholders, and SST sees agency as dispersed focusing on how actors may bring a diversity of perspectives to bear and translate in a design project or innovation process, while the organizing perspective likewise sees agency as dispersed by focus on actions and events as the carriers of agency in contrast to people and has a focus on the world of institutions.

ANT interestingly is a common reference and ontology across the three streams of research that enables shared conceptual understandings and perspectives. The staging perspectives in this book refer to different generations or readings of ANT including different perceptions of relations between action, actors, and networks, as it brings distinctive value to the three traditions. For participatory design ANT adds additional analytical depth, while the shaping technology and organizing approaches are more rooted in ANT but are enriched by the more action-oriented staging approach from participatory design.

Lastly, the theatre metaphor lies behind and provides inspiration for the development of a staging perspective, but while the reference to theatre is explicit in participatory design, and lies as part of the foundation for the organizing tradition, it is only silently present in the shaping of technology approach. However, the focus on spaces enable the metaphor to be a unifying imagery, especially since all the approaches share an interest in the actors involved, invitations, framing, enacting, props, and other aspects that lend themselves well to the theatrical metaphor. And just as ANT lent analytical depth to participatory design, the metaphor makes shaping of technology and organizing more action-oriented.

SPACING AND AGENCY – A CONCEPTUAL FRAME AND REPERTOIRE FOR STAGING

Actors, Objects, and Materiality in Staging

In a theatrical setup staging is guided by a stage director, in negotiation with a host of other actors like writers, choreographers, technicians, and so forth. In participatory design the focus is on the designer, who stages negotiations between multiple actors, which comes closest to the idea of the stage director. The organizing perspective understands actors as actions, taking people out of focus, which is a contrast to participatory design and shaping technology, which, however, brings a stronger focus on what should be done, the action and the performance, and the materiality of knowledge to give it spatial and temporal stability. Drawing upon ANT and the concepts of generalized symmetry also implies that objects and collectives have agency in line with humans, which implies that even an object like a prototype or technology might stage design and innovation if placed in a situation where it makes a difference. This perception of objects, technologies and networks as doing something, even staging particular interactions such as in the case of the internet (Pinch 2010), is essential. So, who is staging? In our perspective staging is an active, overt, and strategic action by one or more actors, human or non-human. Objects can be strategic in the way they are created, the embedded script, the specific framing of discourses, or other ways to stage the interactions an object will participate in, and endow it with agency.

Spaces – Staging and Configuring

While many will connect the theatrical metaphor with a performance happening on a physical stage, we find that design and innovation take place on a range of diverse scenes for negotiation and action. Identifying and/or creating scenes or spaces where something performative might happen from the perspective of design and innovation are key moments in a staging process. Space is a flexible term, and the question is more on the boundaries, i.e. who is invited to the stage and how, and who are merely spectators. The theatre metaphor may further point at the role of potential "invisible workers" preparing and managing aspects of the space.

Generally, we conceptualize spaces as inherently sociotechnical (Clausen and Koch 2002; Clausen and Yoshinaka 2007), but the traditions may emphasize one or more of the relational, political, material, place specific or other aspects. A space can thus be a workshop, a design project, or even a temporarily stable action-net perceived as an organization. An event, however, is a more temporally limited subtype of space. An event is a point in time encompassing interactions between two or more actors, like a meeting or workshop, while even temporary spaces like a design project exist for several events.

Identifying and/or creating spaces where aspects of design and innovation are or may be performed are key moments in a staging process, i.e. situations where choices are or can be made, which is a question of the location and configuration of agency. According to the form of ANT drawn upon there would not be an *a priori* delimitation of a space but rather a follow-up of its elements, drawing a network with probable boundaries. Thus, any such space or socio-material collective would be partly emerging and partly deliberately created.

Having created or identified a space, whatever its form, staging is as in a theatrical setup a guided process – staging is composed of active, overt, and strategic actions by one or more actors. We broadly refer to this as configuring the space, and in the organizational tradition configuration describes a careful consideration of inclusion, enactment, and framing of objects to gain and wield agency (Dorland et al. 2019). Dorland et al. (2019) based on this tradition study specific configurations of actors, researchers, students, communities, spaces, infrastructure, equipment, facilitators, and a host of other entities to assemble agency for different types of activities. The production and circulations of objects such as designed objects, models, prototypes, and documents are key to staging as they carry inscribed knowledge (Akrich 1992) and enable particular interactions and movement of knowledge. This approach considers how to create and maintain a backstage where planning and rehearsals can take place, and actively work with enacting the public in different ways, framing their needs and wants through objects on the scene to further their agenda. Alternatively, in collaborative design settings Pedersen (2020) suggests that designers see themselves as humble stage directors who seek to stage spaces for negotiation of concerns, problems and solutions and for building of compromises rather than pursuing a specific agenda.

The notion of intermediary objects (Vinck 2012) points at the shifting roles objects may perform in terms of representation, mediation, translation, or commissioning, depending on how they are situated and staged. Designers, for instance, typically develop mock-ups, design games and prototypes as "things to think with" as well as objects to circulate as part of strategic facilitation during negotiations between, for instance, designers and users in the hopes that these might work as intermediary objects. In a different context, that Science Shops have statutes limiting clients to a select group and a strategic geographic placement to enable interactions with this group (Dorland et al. 2019) illustrates this point. Such statutes are examples of objects that can be seen as boundary objects (Star and Griesemer 1989) or intermediary objects configured or equipped (Vinck 2011) in such a way that they contribute to an invisible infrastructure allowing both convergence and respected differences.

Both boundary objects and intermediary objects contribute to the repertoire of configuring elements.

Performing and Travelling

The theatre metaphor would suggest that there are several potential outcomes of a performance. The audience may be pleased or at least have had food for thought, the actors may have experienced and learned about the people and scenarios they are trying to portray, and so forth. In a design and innovation setting, the outcome might be knowledge, material objects, new designs, or a changed practice. Also, participatory design focuses on re-framing as a result of heterogeneous negotiations and collective interpretation between actors, objects, business models, etc., within a space, i.e. a democratic aim to empower participants.

Staging efforts can also have a more focused aim, trying to configure networks in particular ways to gain agency. Organizational theorists building on ANT see the emergence of organizations through communication events (Blaschke et al. 2012) based on a process view on organizations drawing on symbolic interactionists like Goffman and Weick (Goffman 1959; Weick 1995). They focus more on the top-down types of staging, for instance by policy makers or executives, in direct contrast to participatory design. Events are in this organizational perspective often staged with a focus on creating objects that can travel and aid the configuration of subsequent events.

In continuation of this, the concept of the intermediary object offered by Vinck (2012) suggests that we follow the object as it moves from site to site and identify the spaces produced by flows and translations of things, people and texts (Vinck 2012: 94). An intermediary object may play a key role in staging the movement of knowledge as it travels from network to network and take on a diversity of roles depending on how it is being circulated and received. The change of current practices by moving knowledge from one space to another has been illustrated by involving city planners in democratic experiments with citizens (Munthe-Kaas and Hoffmann 2017) or by challenging engineering practices by letting engineers participate in collective sensemaking of user practices (Clausen and Gunn 2015).

Repertoires for Action and Insight

The three approaches to staging presented in this chapter each contribute to outlining a first collective understanding of considerations and repertoires concerned with the staging of collaborative design and innovation processes. This is an understanding that future research can build upon. Staging creates collective action in different ways; many strategies are viable, as will be outlined in this book. Thus, the navigational moves related to how to attain power and agency as well as facilitating empowerment are central themes. And the latter is obtained through staging specific configurations of networks that, for instance, create power or focus the dispersed agency. Thus, together the traditions of participatory design, shaping of technology and organizational theory combine some basic considerations, which all attend to agency:

1. **Possible space for staging:** The idea of a space as the focal point where staging plays out is at the core of the repertoire, with considerations of the identification, setup, preparation, delimitation and framing of the space – including who and what to invite, etc. It thus ranges from focusing on creating and preparing the space and framing the interactions to an in-situ staging of the space as one of the occupants.
2. **Role of objects:** Materiality and objects are vital for interactions, and thus for considerations concerning how to identify, select, design, develop, and circulate objects as a central part of staging. Objects can be circulated within smaller or larger networks and thus play a role in configuring agency and stabilizing or destabilizing networks, opening up for change and innovation through their ability to frame interaction and allow knowledge to anchor and travel. How to make knowledge travel over shorter or longer distances and stabilize in the form of designs and innovative solutions is an essential challenge, as well as connecting events over time and space.
3. **Metaphors:** The theatrical metaphor can be seen as a sensitizing device that unifies the spatial metaphors and acts as a unifying imagery that allows for navigational considerations. It should, however, also be noted, that while such metaphors can both be productive and dangerous several chapters in this book illustrate the insights that can be gained.

As staging is a situated process that is to be navigated differently depending on the situation at hand, we hope that this book by offering a number of cases within different areas and with different focus will provide inspiration to a "manyness" of future staging efforts.

REFERENCES

Akrich, M. (1992), 'The de-scription of technical objects', in W.E. Bijker and J. Law (eds), *Shaping Technology / Building Society,* Cambridge, MA: MIT Press, pp. 205–24.

Akrich, M., M. Callon, B. Latour and A. Monaghan (2002), 'The key to success in innovation Part I: The art of interessement', *International Journal of Innovation Management*, **6** (2), 187–206.

Allport, F.H. (1954), 'The structuring of events: Outline of a general theory with applications to psychology', *The Psychological Review*, **61** (5), 281–303.

Andreasen, M.M., C.T. Hansen and P. Cash (2015), *Conceptual Design: Interpretations, Mindset and Models*, New York: Springer International Publishing.
Belliger, A. and D.J. Krieger (2016), *Organizing Networks: An Actor–Network Theory of Organizations*, Bielefeld: transcript Verlag.
Bijker, W.E., T. Hughes and T. Pinch (1987), *The Social Construction of Technological Systems: New Directions in the Sociology and History of Technology*, Cambridge, MA: MIT Press.
Binder, T., E. Brandt, T. Horgen and G. Zack (1998), 'Staging events of collaborative design and learning', in *Proceedings of 5th International Conference on Concurrent Engineering: Advances in Concurrent Engineering*, Tokyo: TMIT, pp. 369–78.
Bjerknes, G., P. Ehn and M. Kyng (1987), *Computers and Democracy – A Scandinavian Challenge*, Aldershot, UK: Avebury.
Björgvinsson, E.B., P. Ehn and P.-A. Hillgren (2010), 'Participatory design and "democratizing innovation "', in *PDC 2010 – Proceedings of the 11th Biennial Participatory Design Conference*, Sydney: Association for Computer Machinery (ACM), pp. 41–50.
Björgvinsson, E.B., P. Ehn and P.-A. Hillgren (2012), 'Agonistic participatory design: Working with marginalised social movements', *CoDesign*, **8** (2–3), 127–44.
Blaschke, S., D. Schoeneborn and D. Seidl (2012), 'Organizations as networks of communication episodes: Turning the network perspective inside out', *Organization Studies*, **33** (7), 879–906.
Boujut, J.-F. and E. Blanco (2003), 'Intermediary objects as a means to foster co-operation in engineering design', *Computer Supported Cooperative Work*, **12** (2), 205–19.
Bødker, S., C. Dindler and O.S. Iversen (2017), 'Tying knots: Participatory infrastructuring at work', *Computer Supported Cooperative Work: CSCW: An International Journal*, **26** (1–2), 245–73.
Brandt, E. (2001), *Event-driven Product Development: Collaboration and Learning*, Lyngby: Technical University of Denmark.
Brandt, E., M. Johansson and J. Messeter (2005), 'The design lab: Re-thinking what to design and how to design', in T. Binder and M. Hellström (eds), *Design Spaces*, Finland: Edita Publishing Ltd IT Press, pp. 34–43.
Brandt, E., J. Messeter and T. Binder (2008), 'Formatting design dialogues – games and participation', *CoDesign*, **4** (1), 51–64.
Bratteteig, T. and J. Gregory (2001), 'Understanding design', in S. Bjørnestad, R.E. Moe, A.I. Mørch and A.L. Opdahl (eds), *Proceedings of the 24th Information Systems Research Seminar in Scandinavia (IRIS 24)*, **3**, Bergen: University of Bergen, accessed 15 May 2020 at http://heim.ifiuio.no/~tone/Publications/Bratt-greg-01.htm.
Bratteteig, T. and I. Wagner (2012), 'Spaces for participatory creativity', *CoDesign*, **8** (2–3), 105–26.
Brodersen, C., C. Dindler and O.S. Iversen (2008), 'Staging imaginative places for participatory prototyping', *CoDesign*, **4** (1), 19–30.
Buur, J. and B. Matthews (2008), 'Participatory Innovation', *International Journal of Innovation Management*, **12** (3), 255–73.
Callon, M. (1986), 'Some elements of a sociology of translation: Domestication of the scallops and the fishermen of St Brieuc Bay', in J. Law (ed.), *Power, Action and Belief: A New Sociology of Knowledge?*, London: Routledge, pp. 196–223.
Callon, M. and B. Latour (2015 [1981]), 'Unscrewing the Big Leviathan: How actors macro-structure reality and how sociologists help them to do so', in K.K. Cetina

and A. Cicourel (eds), *Advances in Social Theory and Methodology: Toward an Integration of Micro and Macro-Sociologies*, London: Routledge & Kegan Paul, pp. 277–303.
Clausen, C. and W. Gunn (2015), 'From the social shaping of technology to the staging of temporary spaces of innovation – a case of participatory innovation', *Science & Technology Studies*, **28** (1), 73–94.
Clausen, C. and C. Koch (2002), 'Spaces and occasions in the social shaping of information technologies: The transformation of IT-systems for manufacturing in a Danish Context', in K.H. Sørensen and R. Williams (eds), *Shaping Technology Guiding Policy: Concepts, Spaces and Tools*, Cheltenham, UK and Northampton, MA, USA: Edward Elgar Publishing, pp. 215–40.
Clausen, C. and Y. Yoshinaka (2005), 'Sociotechnical spaces: Guiding politics, staging design', *International Journal of Technology and Human Interaction*, **1** (3), 44–59.
Clausen, C. and Y. Yoshinaka (2007), 'Staging socio-technical spaces: Translating across boundaries in design', *Journal of Design Research*, **6** (1/2), 61–78.
Czarniawska, B. (2006), 'A golden braid: Allport, Goffman, Weick', *Organization Studies*, **27** (11), 1661–74.
Czarniawska, B. and T. Hernes (2005), *Actor–Network Theory and Organizing*, Malmo: Liber.
Czarniawska-Joerges, B. and G. Sevón (2005), *Global Ideas: How Ideas, Objects and Practices Travel in a Global Economy*, Copenhagen: Copenhagen Business School Press.
Dawson, P., C. Clausen and K.T. Nielsen (2000), 'Political processes in management, organization and the social shaping of technology', *Technology Analysis & Strategic Management*, **12** (1), 5–15.
Dorland, J. (2018), *Transformative Social Innovation Theory: Spaces & Places for Social Change*, Aalborg: Aalborg University.
Dorland, J., C. Clausen and M.S. Jørgensen (2019), 'Space configurations for empowering university–community interactions', *Science and Public Policy*, **17** (5), 689–701.
Ehn, P. (2008), 'Participation in design things', in *PDC '08: Proceedings of the Tenth Anniversary Conference on Participatory Design 2008*, Bloomington: Indiana University, pp. 92–101.
Ehn, P. and D. Sjögren (1991), 'From system descriptions to scripts for action', in J. Greenbaum and M. Kyng (eds), *Design at Work: Cooperative Design of Computer Systems*, Hillsdale: Lawrence Erlbaum, pp. 241–68.
Geels, F.W. and J. Schot (2007), 'Typology of sociotechnical transition pathways', *Research Policy*, **36** (3), 399–417.
Gherardi, S. and D. Nicolini (2005), 'Actor-networks: Ecology and entrepreneurs', in B. Czarniawska-Joerges and T. Hernes (eds), *Actor–Network Theory and Organizing*, Malmo: Liber, pp. 285–307.
Gish, L. and C. Clausen (2013), 'The framing of product ideas in the making: A case study of the development of an energy saving pump', *Technology Analysis and Strategic Management*, **25** (9), 1085–101.
Goffman, E. (1959), *The Presentation of Self in Everyday Life*, New York: Anchor Books.
Hansen, A.G. and C. Clausen (2002), 'The role of TA in the social shaping of technology', in G. Banse, A. Grunwald and M. Rader (eds), *Innovations for an E-Society*, Berlin: Edition Sigma, pp. 91–8.
Hernes, T. (2008), *Understanding Organization as Process. Theory for a Tangled World*, London and New York: Routledge.

Jensen, A.R.V., L. Gish and C. Clausen (2018), 'Three perspectives on managing front end innovation: Process, knowledge and translation', *International Journal of Innovation Management*, **22** (7), 1–32, accessed 17 October 2018 at https://doi.org/10.1142/S1363919618500603.

Jørgensen, M.S., U. Jørgensen and C. Clausen (2009), 'The social shaping approach to technology foresight', *Futures*, **41** (2), 80–6.

Jørgensen, U. and O. Sørensen (2002), 'Arenas of development: A space populated by actor worlds, artefacts and surprises', in K.H. Sørensen and R. Williams (eds), *Social Shaping of Technology: Concepts, Spaces and Tools*, Cheltenham, UK and Northampton, MA, USA: Edward Elgar Publishing, pp. 197–222.

Kamp, A. (2000), 'Breaking up old marriages: The political process of change and continuity at work', *Technology Analysis & Strategic Management*, **12** (1), 75–90.

Kimbell, L. (2011), 'Rethinking Design Thinking: Part I', *Design and Culture*, **3** (3), 285–306.

Kjølner, T. (2007), 'Iscenesættelse', Gyldendals Teaterleksikon, accessed 25 July 2019 at http://denstoredanske.dk/Gyldendals_Teaterleksikon/Begreber/iscenesættelse.

Latour, B. (1987), *Science in Action: How to Follow Scientists and Engineers through Society*, Cambridge, MA: Harvard University Press.

Law, J. (2002), 'Objects and spaces', *Theory, Culture & Society*, **19** (5/6), 91–105.

Law, J. and K. Hetherington (2000), 'Materialities, spatialities, globalities', in J.R. Bryson, P.W. Daniels, N. Henry and J. Pollard (eds), *Knowledge, Space, Economy*, London, New York: Routledge, pp. 34–49.

Michael, M. (2017), *Actor–Network Theory: Trials, Trails and Translations*, Thousand Oaks, CA: Sage Publications.

Micheli, P., S.J.S. Wilner, S.H. Bhatti, M. Mura and M.B. Beverland (2018), 'Doing Design Thinking: Conceptual review, synthesis, and research agenda', *Journal of Product Innovation Management*, **36** (2), 124–48.

Molina, A.H. (1997), 'Insights into the nature of technology diffusion and implementation: The perspective of sociotechnical alignment', *Technovation*, **17** (11–12), 601–26.

Munthe-Kaas, P. and B. Hoffmann (2017), 'Democratic design experiments in urban planning – navigational practices and compositionist design', *CoDesign*, **13** (4), 287–301.

Pedersen, S. (2020), 'Staging negotiation spaces: A co-design framework', *Design Studies*, **68**, 58–81.

Pel, B., J. Dorland, J. Wittmayer and M.S. Jørgensen (2017), 'Detecting social innovation agency', *European Public & Social Innovation Review (EPSIR)*, **2** (1), 1–17.

Pinch, T. (2010), 'The invisible technologies of Goffman's sociology: From the merry-go-round to the Internet', *Technology and Culture*, **51** (2), 409–24.

Rip, A. and J. Schot (2002), *Identifying Loci for Influencing the Dynamics of Technological Development*, Cheltenham, UK and Northampton, MA, USA: Edward Elgar Publishing.

Robichaud, D. and F. Cooren (2013), *Organization and Organizing : Materiality, Agency, and Discourse*, London and New York: Routledge.

Russell, S. and R. Williams (2002), 'Social shaping of technology: Frameworks, findings and implications for policy', in K.H. Sørensen and R. Williams (eds), *Shaping Technology, Guiding Policy: Concepts, Spaces and Tools*, Cheltenham, UK and Northampton, MA, USA: Edward Elgar Publishing, pp. 37–131.

Simonsen, J. and T. Robertson (2012), *Routledge International Handbook of Participatory Design*, London and New York: Routledge.

Sørensen, K.H. and R. Williams (2002), *Social Shaping of Technology: Concepts, Spaces and Tools*, Cheltenham, UK and Northampton, MA, USA: Edward Elgar Publishing.
Star, S.L. and J.R. Griesemer (1989), 'Institutional ecology, "translations" and boundary objects: Amateurs and professionals in Berkeley's Museum of Vertebrate Zoology, 1907–39', *Social Studies of Science*, **19** (3), 387–420.
Storni, C., T. Binder, P. Linde and D. Stuedahl (2015), 'Designing things together: intersections of co-design and actor–network theory', *CoDesign*, **11** (3–4), 149–51.
Storni, C., T. Binder, D. Stuedahl and P. Linde (2012), 'Exploring ANT in PD: Reflections and implications for theory and practice', in *Workshop at Participatory Design Conference 2012*, Roskilde, Denmark: PDC, pp. 1–47.
Strum, S.S. and B. Latour (1987), 'Redefining the social link: From baboons to humans', *Social Science Information*, **26** (4), 783–802.
Stuedahl, D. and O. Smørdal (2015), 'Matters of becoming, experimental zones for making museums public with social media', *CoDesign*, **11** (3–4), 193–207.
Vinck, D. (2011), 'Taking intermediary objects and equipping work into account in the study of engineering practices', *Engineering Studies*, **3** (1), 25–44.
Vinck, D. (2012), 'Accessing material culture by following intermediary objects', in L. Naidoo (ed.), *An Ethnography of Global Landscapes and Corridors*, Rijeka, Croatia: InTech, pp. 89–108.
Weick, K.E. (1995), *Sensemaking in Organizations*, Thousand Oaks, CA: Sage Publications.

PART II

Staging participatory co-design with multiple actors

3. Staging negotiation spaces as a means for co-designing an insulin service system in India

Signe Pedersen and Søsser Brodersen

INTRODUCTION

Diabetes is a global challenge; currently, it is the fastest growing disease in India. According to the World Health Organization (WHO), 69.2 million people in India were living with type 2 diabetes in 2015, and that number is expected to grow to 98 million people by 2030. Over the next 12 years, 20 percent more insulin will be needed to effectively treat type 2 diabetes worldwide (Basu et al. 2018). Clearly, 'substantial improvements in access to insulin in low-income and middle-income countries are needed in order to reduce inequalities in access and complications of diabetes compared with high-income countries' (Basu et al. 2018: 2).

To address this challenge, one of the world's leading pharmaceutical companies established a cross-sectional sub-division focused on base of the pyramid (BoP) markets, and initiated a project with the aim of investigating how to 'bring insulin closer to the poorer patients with diabetes in India'. The usual design approach of this Denmark-based company is focused on developing products for the Western market such as insulin drugs and insulin pens for injection purposes. This research and development (R&D) process is traditionally driven from the top-down, and revolves around collaboration with highly influential experts such as doctors affiliated with major hospitals around the world. However, this new sub-division was given the freedom to initiate projects of an explorative nature from the bottom-up. Because the sub-division had very little knowledge about the BoP segment in India or the local conditions, it adopted an explorative approach to investigate and understand the concerns of poor people living with diabetes in India. In other words, the sub-division was allowed to explore a market where the value creation process typically involves a high degree of uncertainty (Cholez et al. 2012).

Typically, companies enter the BoP market because they view end-users as potential consumers who lack access to new products and services, or as producers and entrepreneurs with the potential to become actors in future value chains (Cholez et al. 2012). In line with the user-as-consumer approach, the focus in this project was to investigate how an already developed insulin drug could be made accessible and available in unserved parts of India, such as the New Delhi slums. This is a typical approach when entering in emerging markets, as a central point in this process is not necessarily to develop new products, but to design and develop an ecosystem of central actors such as companies, non-governmental organizations (NGOs), micro-entrepreneurs, public sector stakeholders and others to ensure the availability of existing products (Prahalad 2009; Reficco and Márquez 2012). Although the BoP literature provides many interesting examples of innovative ecosystems, few studies describe how to build new and innovative network constellations to serve emerging markets.

The company struggled to initiate this project, and therefore decided to hire two external design engineers to help them investigate the local conditions in India. The two external designers come from a participatory design background and thus framed the project in a collaborative manner using a staging approach to engage central actors from a potential future ecosystem of people with diabetes (PwDs), doctors, pharmacists, NGOs, micro-entrepreneurs, public hospitals and local health practitioners.

In this chapter, we describe the endeavours of the two design engineers (from here on termed 'the designers'). Viewing these designers as stage directors who staged negotiations among multiple actors reveals insights about how to navigate complex design projects involving multiple actors. To further explore the navigational role of the designers, we draw upon the analytical and normative framework of staging negotiation spaces (Pedersen 2020) to discuss how staging contributes to understanding how to build and navigate networks and systems in collaborative design projects involving multiple actors and concerns.

WHAT CAN BE LEARNED FROM STAGING?

The staging negotiation spaces framework is inspired by the concept of theatrical 'staging': designers are viewed as empathic and explorative stage directors who stage discursive spaces for actors to negotiate matters of concern. Here, the metaphor of staging enables us to draw on useful concepts from actor–network theory (ANT) and participatory design to analyse the negotiations taking place during complex design and innovation activities.

Although the BoP literature points to the need to build ecosystems, few insights into how this is actually done are available. Drawing inspiration from ANT, participatory design scholars and practitioners have developed an actionable approach to building infrastructures and networks. In this view, designers are not seen as merely creative experts but as stagers and facilitators of encounters between users and other actors (Iversen et al. 2012) whose relations and concerns are often conflicting and contested (Binder et al. 2015; Björgvinsson et al. 2012). Thus, staging is the act of building network relations by orchestrating spaces for the negotiation of concerns (Pedersen 2020).

The theatre metaphor enables us to think about the elements involved in the staging process; in the Scandinavian theatre tradition, staging is a process conducted by a stage director who interprets a text, frames the play based on this interpretation and eventually configures the scenic space by casting actors and visualizing props in the form of objects (Kjølner 2007). In participatory design, this process corresponds to designers (a) interpreting a design brief, (b) framing negotiations between actors based on the interpretation, and (c) configuring a discursive space by developing props and inviting actors at many levels to participate. Examples of props include design games, mock-ups and prototypes which designers inscribe with a 'framework of action' based on their framing of the space (Akrich 1992). The idea is for the designers to continuously bring materializations of their framings into circulation amongst the actors involved in the hopes that they can serve as intermediary objects (Boujut and Blanco 2003; Vinck 2012) in negotiations. The concept of intermediary objects describes objects' ability to *represent*, *translate* and *mediate* between actors (Vinck 2012) and help us understand how objects influence negotiations between other actors, which is a central part of building networks through translation.

Staging may apply to networks of different sizes, which means that the 'performance' being staged can take different forms depending on the analysis. For instance, authors such as Hoffmann and Munthe-Kaas (Chapter 14, this volume) explain how a new area of a city is developed by framing the new area as a performance which needs to be staged. This view implies that the desired performance of the final network is already known. However, in this chapter, we focus on the staging of multiple, sequential negotiation spaces during a design project wherein the network spanning the final system or solution was gradually negotiated and built through conceptualization. This also means that, in contrast to a theatre play with a fully planned scenic performance including rehearsals and timetables, the performances in question in this chapter rather resemble improvisation, whereby the performance is continuously negotiated and enacted in a series of one-act performances.

From this perspective, the performances discussed in this chapter are those of negotiations between actors comprising the 'final' performance of bringing insulin closer to the patients in India. We see the negotiations between human as well as non-human actors as central in mobilizing actors to play a role in the network being designed—in this case, a system that increases access to insulin among poor diabetes patients in India.

In the case analysis, we use the staging negotiation spaces framework to reveal the navigational moves of the designers in their efforts to stage negotiations among multiple actors and build a network to improve healthcare in emerging markets by: (a) framing an envisaged performance in a scenic space based on an interpretation of the initial 'design brief' of a project to bring insulin closer to PwDs; (b) producing props that inscribe this framing into potential intermediary objects in the form of scenario cards and value cards; (c) circulating these objects amongst diverse actors to facilitate negotiations; and (d) reframing the design brief to stage yet another performance with new actors and props.

METHODOLOGY

In the following case, we illustrate how two designers staged, facilitated and reframed spaces for negotiations in India and in Denmark with the aim of negotiating matters of concern and translating knowledge. One of the authors of this chapter was assigned to the project for a period of three months and participated in the development of scenario cards, meetings with the company project manager, and ten days of field studies in India. During the fieldwork, the designer conducted interviews, participated in informal meetings, engaged in network-building activities, facilitated design games, and observed the activities of 20 PwDs and eight 'experts'. The author also participated in data analysis and synthesis, concept development and the handover of the project to the company.

Different sites, including a specialized diabetes hospital in India, households in slum areas of New Delhi, and office buildings hosting NGOs and entrepreneurial companies, provided the backdrop for the fieldwork. The work was guided by traditional ethnographic field study techniques (van Maanen 2011), including observations and qualitative interviews (DiCicco-Bloom and Crabtree 2006) as the researchers 'followed the actors' (Latour 1987). The first author kept a diary and transcribed interviews, which together with design objects such as value cards, scenario cards, PowerPoint presentations, and others became the basis for the analytical research process. With respect to the data analysis, the case narrative was written based on an abductive analysis (Timmermans and Tavory 2012) carried out by both authors in reflective discussions to ensure objectivity.

The structure of the case study is organized according to what can be seen as three different negotiation spaces. Using the staging negotiation spaces framework, we present an in-depth analysis of the iterative nature of staging in the first space, which involved multiple episodes of facilitating negotiations and reframing. The first author played an active role in these staging efforts. The next two spaces are not analysed to the same depth, but serve as examples of navigating negotiation spaces with very different configurations and framings.

DESIGNING AN INSULIN SYSTEM IN INDIA

According to the design brief provided by the company, the objective of the project was to develop a scalable business model to increase access to insulin among poor PwDs in India. At this time, projects in emerging markets were framed by the company's marketing department and typically revolved around sales opportunities. The only contacts the company initially had in India were personnel in its sales offices and its customers, which were limited to major hospitals with the capacity to treat diabetes. Typically, general practitioners were involved as central actors in the company's market expansion efforts, because they were the ones prescribing the insulin; however, this was not the case in India.

Negotiation Space 1: Negotiating PwDs' Concerns in India

One of the designers had previously worked as an external consultant for the company on a preliminary study involving students from a major university in Denmark. Therefore, she already understood the purpose of the company's new exploratory sub-division and its desire to target the BoP segment in emerging markets by introducing sustainable and scalable business models. Thus, her interpretation of the design brief was to a large extent aligned with the company's intentions. However, the project framing was also influenced by the other designer and their common background as participatory designers. The usual approach of the company would be to interview experts such as general practitioners (GPs) to understand how to provide better solutions for patients. However, because the designers had years of experience co-designing with users, governments, entrepreneurs, and larger companies in emerging markets, they strived to actively engage not only GPs, but a wide range of actors in a creative dialogue. In line with ANT's vocabulary indicating that design and innovation are essentially about network building, the designers were very aware that in order to increase PwDs' access to insulin, they would need to design a network to span the final solution. Thus, they found it important to identify any actors who would potentially play a part in the network, and to engage as many of them as possible. As a result, they invited PwDs, NGO workers, entrepreneurs, and professionals to participate in negotiations and network-building activities. Therefore, they slightly reframed the company's design brief to involve more actors than 'the usual suspects' to understand the concerns of PwDs in India.

The two designers also felt it was important for non-human actors to participate in the negotiations. Thus, they developed two ranking-based design games to spark conversation with PwDs as well as others. One game involving a set of value cards was targeted specifically to PwDs, and another set of scenario cards was designed to facilitate engagement with a broader range of actors.

Inscribing the company's values into value cards

The value cards were intended as a way to initiate conversation and negotiation with PwDs about the set of articulated values that typically guided the company's actions. The company was completely immersed in the global health paradigm, and thus worked with key concepts promoted by the WHO and United Nations High Commissioner for Human Rights, such as availability, accessibility, and acceptability. As the company generally strove to live up to some of these 'basic rights to health' it also used these concepts to evaluate potential new projects and solutions. From the company's perspective, it was important for the designers to explore how diabetes patients in India perceived their situation relative to these key concepts. Furthermore, the company was interested in knowing about the level of affordability and whether PwDs were interested in receiving home-based services such as help with insulin injections. The designers interpreted the company's desire to understand PwDs' thoughts on these key concepts and designed six value cards reflecting these

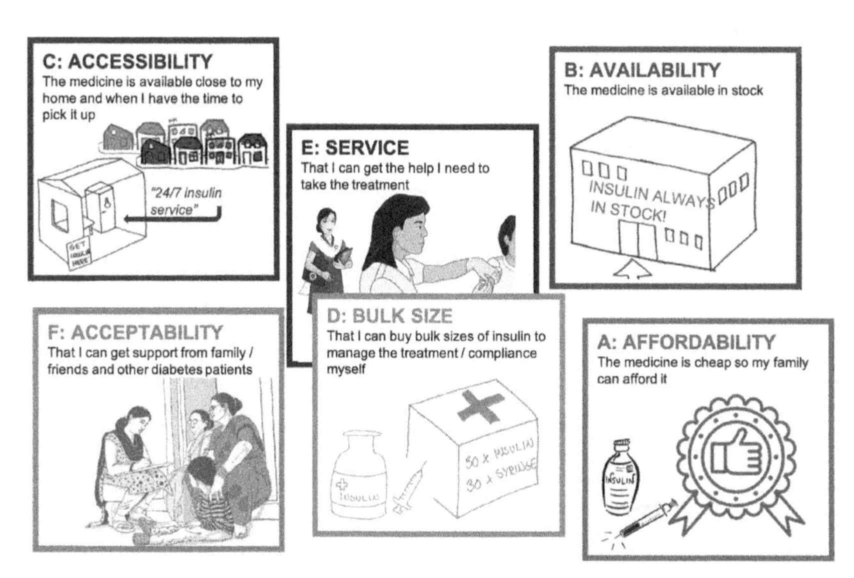

Figure 3.1 *Value cards to be used in engagements with key actors in India*

value concepts (see Figure 3.1). They also brought with them a number of blank cards that could be filled in 'on location' should new concerns arise during the negotiations.

Inscribing scenario cards as provotypes

The two designers also wanted to create objects to frame the dialogue and open up discussion about additional aspects of life with diabetes in India. Their interpretation of the original design brief can be viewed as reflecting their desire to come up with solutions to bring insulin closer to the patients. They brainstormed and came up with six different scenarios illustrating different solutions to increase PwDs' access to insulin in India (see Figure 3.2 for an illustration of four of these). These depicted numerous general ideas, such as having the insulin available at the local hospital, being able to buy insulin at a local establishment, selling the insulin from a 'pharmacy on wheels', and having a local health worker visit patients at home to help them inject the insulin. The scenario cards represented the explorative and collaborative framing as well as the previous experiences of the designers, both of whom had designed solutions for emerging markets and associated cultures.

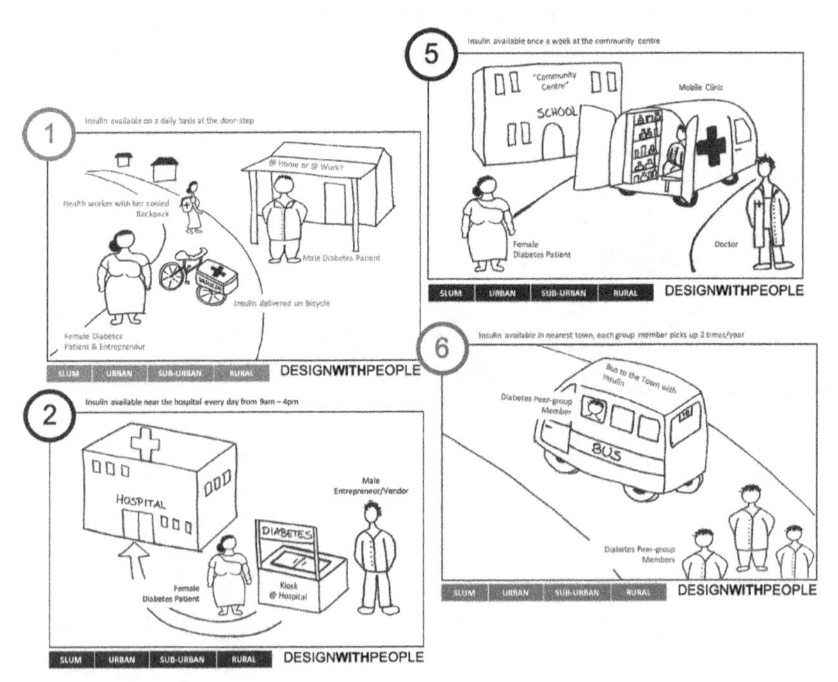

Figure 3.2 *Scenario cards to be used in engagements with key actors in India*

The scenarios were constructed as provotypes (Boer et al. 2013) aiming to test as many different aspects of the scenarios as possible while creating opportunities for the different actors to critique the ideas. The designers inscribed provocative elements in some of the scenarios based on their previous experiences engaging with people from highly hierarchical cultures in emerging markets, where especially people with lower status might not initially feel comfortable critiquing foreign visitors. For instance, in some scenarios, the providers of insulin were male and in others female, some were doctors or pharmacists, while others were local entrepreneurs. In one scenario, a female entrepreneur was biking, which is potentially controversial in certain religious cultures. Such elements provided opportunities for PwDs to educate the designers on their own culture rather than simply saying that the concept was a bad idea.

Together, the two sets of cards represented the company's concerns and the designers' experiences and ideas. Both the value cards and the scenario cards were colourful to spark interest, and the pocket-size format enabled them to be easily transported. Considering the existing physical arrangement of the space they were staging, the designers printed and glued the scenario cards to thick cardboard so they would not fly around when exposed to fans used to cool the air in the small houses in India. The illustrations were rendered in a low fidelity, cartoon-like format to highlight the fact that these were only initial ideas that were open for critique. Furthermore, the women in the drawings wore saris to illustrate that these scenarios were developed specifically for the context. This was done to ease the descripting (Akrich 1992) burden for the local population.

The designers also produced a timetable indicating planned meetings with actors such as NGO workers and managers, GPs and entrepreneurs. Some of these were already members of the designers' existing network in India, whereas others were found through simple Google searches on the internet; all were invited to participate in co-designing this new network comprising PwDs, GPs, insulin, refrigerators, etc.

To ensure that their interpretation and framing of the design brief aligned with the company's expectations, the designers invited a company representative to a meeting wherein they presented the value cards, scenario cards, and their timetable, indicating which people they planned to meet. The project manager immediately approved the cards and the timetable, establishing alignment between the designers and the company before their departure to India.

Inviting PwDs and rehearsing the performance

Upon their arrival in India, the two designers met with several doctors, NGO workers and a small company that had developed an innovative business model to deliver eye tests and glasses to poor people living in rural areas. To accomplish the second part of the staging—'casting' the PwDs who would represent the end-users—the designers met with the local NGOs and asked them to arrange individual meetings with representative local PwDs and their loved ones in their private homes in the New Delhi slum areas. They decided to set up negotiation spaces in PwDs' own homes to create a sense of comfort and freedom to express themselves. Also, the designers were aware of the social status hierarchy in India, and did not wish to create spaces with uneven power relations between the actors.

During a meeting with a local GP who also taught at a medical school in New Delhi, the designers were introduced to a student who could serve as a translator when engaging with PwDs. The designers cast this student as a central actor in the negotiation space with diabetes patients and asked him to translate the card texts into Hindi to ease understanding and minimize the burden of descripting for PwDs, and to facilitate a sense of ownership and comprehension of the aim of the sessions. The interpreter was instructed by the two designers on how to use the cards to facilitate negotiations with the PwDs. By playing a central role in the space, the student enabled the designers to take a step back and observe the negotiations. Drawing on the theatre metaphor, this process may be seen as a rehearsal of the forthcoming negotiations with the local PwDs. The design games not only were literally translated by the medical student, but also served as intermediary objects that translated the project brief and potential solutions, and enabled the student himself to become familiar with the project and prepare for the performance that was to take place the next day.

Facilitating negotiations of matters of concern

> The air is hot and the fan is gushing cooler wind through the small room where eight people are gathered: two designers, the interpreter, a PwD who is the inhabitant of the house, as well as four other family members who are intrigued by the presence of foreign Danish designers and a medical student and want to see what all the fuss is about. The walls of the house are painted in a bright pink colour, and the carpet on the floor is woven with intricate patterns and colours. All eight people are sitting on the floor with the cards spread out between them. One by one, the interpreter asks the PwD how he feels about the current availability and affordability of insulin and whether he would like someone to come and help him with his daily injections. Once they have gone through all of the value cards, he is asked to rank the cards in terms of what is most important to him. He moves the cards around for a few moments and then pauses. He asks: 'What about the quality of the medicine?' The interpreter translates into English for the designers to respond, but they do not have an answer to this. Instead they ask the interpreter to ask what the patient means, and he explains that they experience huge problems with counterfeit medicine in India. So, the most important thing for him would be to know for certain that the medicine is of good quality. The two designers thank him for this great insight and immediately begin drawing and writing 'Quality of Medicine' on one of the blank cards

> they have brought with them. The card immediately becomes part of the ranking process and is ranked number one in terms of importance.

The vignette illustrates how the PwD reconfigured the space by introducing a new value to be ranked. The designers quickly translated this request and immediately prepared a new value card inscribing the patient's quality concern. This card was added to the existing set to be used in subsequent spaces with other PwDs. The designers' ability to navigate and reconfigure the space instantly changed their role from spectators of the 'performance' (that is, the negotiations among the PwD, the interpreter acting as facilitator and the value cards) to become actors in the performance by interpreting and inscribing the articulated concerns of the PwD into a new prop—the quality value card. Whereas the rest of the value cards represented the values of the WHO and thus of the company, the quality card represented the main value of a PwD in India and became an intermediary object that not only mediated among the designers, interpreter and PwD in this space, but also had the potential to travel and become an intermediary object in subsequent spaces. Importantly such travel is only viable if an object represents a stable network that can be recognized by other actors. Thus, the result of the negotiation in this space was not only a new value card (which ultimately turned out to be the highest ranked card for all PwDs involved), but also a slight reframing of the initial concern of bringing insulin closer to patients to bringing *identifiable quality medicine* closer to patients.

Navigating contradictory concerns

From the general feedback on the scenario cards, it was interesting to observe the range of different cultural conceptions attached. For instance, whereas some experts claimed that 'women don't bike around in the streets', others stated 'of course women bike—it is a progressive statement to do so'. The common denominator here was a concern about females biking *alone*. Both experts and PwDs expressed concern for women who biked around alone with no male family members to take care of them. Several suggested that a man and a woman could go together, thereby also ensuring coverage of a broader area. Others mentioned that in certain areas, women were not allowed to treat men: 'The injection should happen in the arm or in the thigh—and especially the thigh is a very private place. This means, that a female entrepreneur in the rural setting cannot treat a male patient. It should be a male health worker for men' (Female board member for a patient group in India).

The designers learned a lot about life in India and particularly the lives lived by residents with diabetes. While listening to the PwDs' stories as they engaged with the different cards (particularly the scenario cards), a translation process occurred. The cards no longer represented only the initial ideas and framings of the designers, but the concerns and lives of the PwDs, the concerns of the NGO workers, and so forth. Representing the views of many different patients and other actors naturally also brought many contradictory insights to the surface. Bringing quality medicine closer to patients is indeed a 'matter of concern' in the Latourian sense, meaning that it is complex and contested. The designers interpreted these conflicting statements and framed the design space to encompas a flexible system that could be adapted to different cultures and areas (see Figure 3.3).

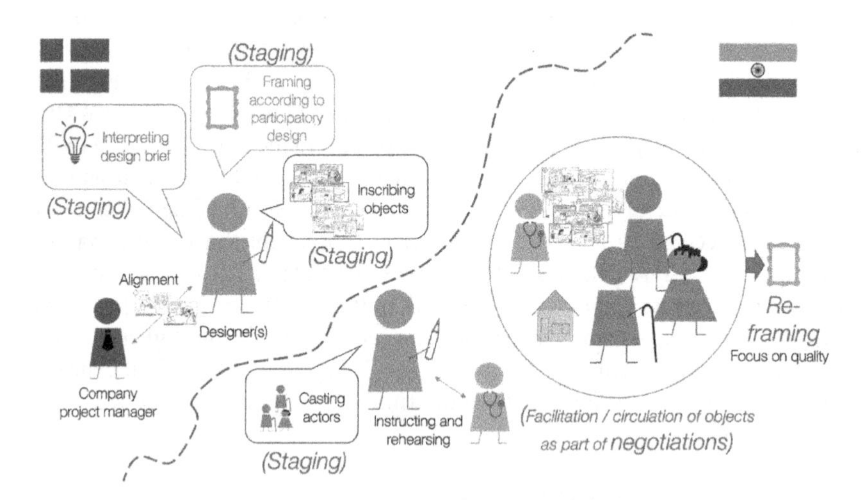

Figure 3.3 *Negotiation space # 1: staging negotiations with PwDs across continents*

Negotiation Space 2: Staging Negations with the Company in Denmark

Back in Denmark, the designers began to make sense of the empirical material they had collected in India. First, they prepared an affinity diagram to make sense of all of the quotes, notes and pictures they had collected. They inscribed their learnings and negotiated with the data, trying to group statements and insights. Eventually, they assigned labels to clusters of information reflecting the central theme of each group, including 'quality is central' and 'qualifications of the entrepreneurs', under which a sub-theme emerged called 'address gender norms'. This affinity diagram represented the many people who had been engaged in India as well as the designers' interpretation of their concerns. The contact person from the company had shown examples of previous deliverables from external consultants, all of which had been created using PowerPoint or similar presentation tools.

Thus, the designers translated and inscribed the concerns that had emerged as part of this analysis into a PowerPoint presentation which, together with pictures and maps from India, formed the materiality of the 'formal' handover to the company. They did this in the hopes that the PowerPoint would serve as a central intermediary object in the negotiation space with the company representative based on previous experiences handing over similar projects. Another central element was a physical illustration on paper of a flexible and scalable solution to bring quality insulin closer to the patients. This illustration depicted a network spanning the solution and indicated the relations between human and non-human actors across blocks, towns and villages which the designers had developed during an unplanned brainstorming session. Even though the solution was not explicitly requested by the company, which had only asked for insights into the concerns of PwDs in India, the designers decided to bring it to the negotiation space as part of the delivery.

As part of facilitating the negotiation space, the designers invited the project manager into the private home of one of the designers for an informal coffee meeting. They presented the affinity diagram, which was laid out and covered the entire dining room table, the PowerPoint deck, and the illustration of the proposed network (product–service system) to the project manager. Narratives illustrating the stories and concerns of the people they had met in India comprised a central part of the presentation. The designers emphasized that quality was mentioned as a central value and that collectively, the PwDs had ranked it as the most important, even though it had not been available for ranking in the first couple of interviews. Surprisingly, the project manager was easily convinced, perhaps because the visible data and insights clearly represented a vast amount of new information about the conditions in India. The props in this negotiation space rendered the work that the designers had done visible, and tangibly represented the concerns of all of the invited actors spanning the network-based solution.

In this case, the negotiation space was like a theatre play in which the project manager representing the company was not so much a participant in negotiations, but more of an engaged spectator of a show performed by the designers, their PowerPoint presentation, and the visual representation of the solution. This particular spectator obviously enjoyed the performance as she asked the designers to translate these insights into something she could present to the decision makers at the company to initiate a pilot project to test the concept areas (see Figure 3.4).

Negotiation Space 3: Influencing the Decision-making Negotiation Space

In contrast to the first two spaces, the third and final negotiation space presented in this chapter is an example of a space over which the designers had

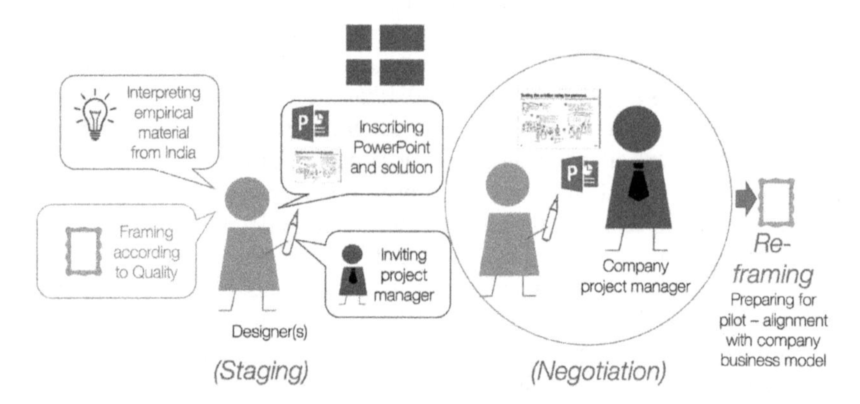

Figure 3.4 *Negotiation space # 2: staging negotiations with company project manager*

little influence in terms of framing and configuration. Because this negotiation space was a 'gate' in the company's stage–gate model, the company's internal business procedures as well as the stage–gate model itself highly framed and configured this space. The configuring elements were decision makers, the project manager, and a standard PowerPoint template inscribed with specific headlines to present the business case. These headlines asked for inputs on, for example, description and business impact, resources, costs, key performance indicators (KPIs), and risk, and only allowed for a few lines of text and a couple of bullet points for each topic to present the potential of the new concept. Thus, in this space, the affinity diagram, the PowerPoint presentation with key insights from India, and the illustration of the network spanning the product–service system were excluded.

The designers were frustrated and struggled to translate the 76 pages of rich empirical data into a template with predefined headlines and only room for a few bullet points for each topic. This template forced the designers to select only a few central aspects of the business case of the proposed concept. Nevertheless, together with the designers, the project manager, who was familiar with the company's decision-making processes and able to navigate them, counter-staged the space by stretching the format of the evaluation template a bit. Rather than only presenting the business potential of the new solution (which was the aim of the template) the designers added more slides to the presentation and hoped that the company decision makers would allow this counter-staging. Thus, the designers and the project manager negotiated the tight framing by preparing what they considered to be the most relevant aspects of the insights and eventually ended up producing a ten-page presentation (including a one-page template that was filled in with the required data) for the project manager to present at the meeting. The project manager rehearsed her presentation and presented the concept and business case to decision makers. The audience (the decision makers) accepted the reconfiguration of the space in the form of the extended PowerPoint presentation, as it conformed to the template just enough to satisfy their expectations. The presentation thus managed to serve as an intermediary object representing the data, the PwDs in India, the final concept, *and* the business case that the decision makers wished to see, thereby translating and mediating among the actors in India, the project manager, and the decision makers at the company (see Figure 3.5).

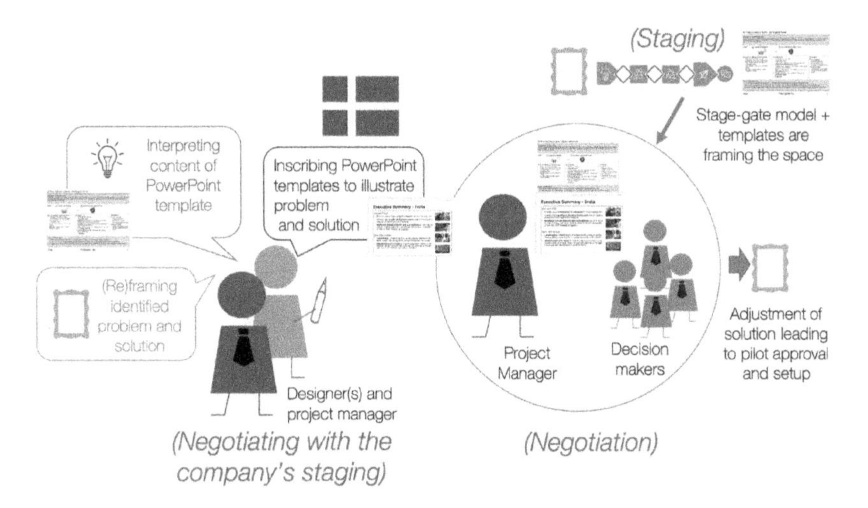

Figure 3.5 *Negotiation space # 3: staging negotiations with decision makers*

Neither of the authors of this chapter know the exact details about the negotiations that took place in this third negotiation space involving the decision makers; however, we do know that it resulted in the approval of a pilot study in India. Sticking with the theatre metaphor, this pilot study was staged as a rehearsal of a product–service system to bring insulin closer to PwDs in India. This rehearsal was applauded by spectators in the form of PMGroup, a respected independent publisher and business information provider based in the UK. PMGroup awarded the project a 'Pharmaceutical Market Excellence Award' in 2013, for its innovative provision of patient and home-care services for the BoP market.

DISCUSSION AND CONCLUDING REMARKS

The staging negotiation spaces framework encourages us to explore design and innovation as a series of temporary negotiations with multiple actors. Viewing the design process as a series of iterative cycles breaks with the perception of design as a linear process that can be planned in advance. Instead, the designer reacts to outcomes of negotiations in the previous space rather than following a master plan (script) to the letter. Whereas the traditional theatrical metaphor celebrates the final performance on stage as resulting from adherence to timetables and scripts throughout numerous rehearsals, the staging negotiation spaces approach resembles that of improvisational theatre as a series of multiple one-act performances, and thus points to the iterative nature of staging numerous negotiations during the design process. The network is the result of these negotiations: the system of bringing insulin closer to PwDs in India could not be staged until collaborative negotiations had taken place.

In this chapter, staging is understood as the interpretational and framing actions involved in configuring a negotiation space wherein concerns are negotiated and produce translations and reframings. The designers engaged numerous actors throughout the design process by continuously staging spaces for negotiation, thereby setting the stage for the translation of actors into a network. In this case, co-designing a scalable business model involved staging negotiations not only with actors in India, but also with company representatives in Denmark, as actors in the possible network of the new solution negotiated their roles and identities along the way (Akrich et al. 2002). Depending on the space, designers have more or less freedom to frame and configure it. For instance, the designers had higher degrees of freedom at the beginning of the project than at the end, when the company stage–gate model and decision-making process framed the space extremely tightly. Thus, a staging perspective helps bring to light the navigational possibilities and moves of the designer(s) before and between the staged events.

We also see how staging is often a collaborative effort. The two designers negotiated the staging throughout the design process and also involved actors such as the project manager to ensure alignment, the NGO directors to help recruit PwDs, and so forth. This leads us to conclude that staging might occur over time and even continents, as some aspects of staging such as the interpretation of the project brief, the inscription of value and scenario cards as well as the initial identification of interesting actors took place in Denmark, whereas others, such as the recruitment of PwDs, the instruction of the interpreter, and the rehearsal of negotiations were conducted in India. Rehearsal and instructions are especially required when different people play the roles of stager and facilitator/circulator of objects in the negotiations. As was the case in the first and third spaces, the designers participated in staging efforts whereas the circulation of objects was primarily handled by the interpreter and the project manager, respectively.

The case also reveals how the scripts of circulated objects frame negotiations and can be understood as intermediary objects (Vinck 2012). For instance, during the negotiations in the first space, the same objects were translated from representing predefined ideas and values to representing the concerns of the patients with diabetes. Moreover, through the staging efforts of the designers in terms of interpreting the empirical material, these value cards and scenario cards were translated into the presentation and the final solution used to establish alignment with the company project manager in the second negotiation space. Thus, we see how materiality and knowledge travels from one space to the next if it represents a stable network. The circulation of objects in as well as between spaces plays a central role in network-building efforts as it enables actors with contradictory concerns to negotiate and communicate (for instance on the topic of biking), thereby contributing to a flexible and scalable network and business model (Binder et al. 2015). To sum up collaborative designers must possess exceptional navigational skills in terms of (a) framing envisaged performances in a scenic space based on continuous (re)interpretation of the initial 'design brief'; (b) producing props that inscribe this framing into potential intermediary objects; (c) circulating these objects amongst diverse actors to facilitate negotiations of concerns; and (d) reframing the design brief to stage yet another performance with new actors and props. Importantly, these skills are needed particularly with interpretational and actionable aspects of staging, to build new networks spanning tomorrow's solutions.

REFERENCES

Akrich, M. (1992), 'The de-scription of technical objects', in W.E. Bijker and J. Law (eds), *Shaping Technology / Building Society: Studies in Sociotechnical Change*, Cambridge, MA: MIT Press, pp. 205–24.

Akrich, M., M. Callon and B. Latour (2002), 'The key to success in innovation Part I: the art of interessement', *International Journal of Innovation Management*, **6** (2), 187–206.

Basu, S., J.S. Yudkin, S. Kehlenbrink, J.I. Davies, S.H. Wild, K.J. Lipska, J.B. Sussman and D. Beran (2018), 'Estimation of global insulin use for type 2 diabetes, 2018–30: A microsimulation analysis', *The Lancet Diabetes and Endocrinology*, **7** (1), 25–33.

Binder, T., E. Brandt, P. Ehn and J. Halse (2015), 'Democratic design experiments: Between parliament and laboratory', *CoDesign*, **11** (3–4), 152–65.

Björgvinsson, E.B., P. Ehn and P.-A. Hillgren (2012), 'Agonistic participatory design: Working with marginalised social movements', *CoDesign*, **8** (2–3), 127–44.

Boer, L., J. Donovan and J. Buur (2013), 'Challenging industry conceptions with provotypes', *CoDesign*, **9** (2), 73–89.

Boujut, J.-F. and E. Blanco (2003), 'Intermediary objects as a means to foster co-operation', *Engineering Design Computer Supported Cooperative Work*, **12** (2), 205–19.
Cholez, C., P. Trompette, D. Vinck and T. Reverdy (2012), 'Bridging access to electricity through BOP markets: Between economic equations and political configurations', *Review of Policy Research*, **29** (6), 713–32.
DiCicco-Bloom, B. and B.F. Crabtree (2006), 'The qualitative research interview', *Medical Education*, **40** (4), 314–21.
Iversen, O.S., K. Halskov and T.W. Leong (2012), 'Values-led participatory design', *CoDesign*, **8** (2–3), 87–103.
Kjølner, T (2007), 'Iscenesættelse', Gyldendals Teaterleksikon, accessed 25 July 2019 at https://teaterleksikon.lex.dk/iscenesættelse.
Latour, B. (1987), *Science in Action: How to Follow Scientists and Engineers through Society*, Cambridge, MA: Harvard University Press.
Pedersen, S. (2020), 'Staging negotiation spaces—a design framework', *Design Studies*, **68**, 58–81.
Prahalad, C. K. (2009), *The Fortune at the Bottom of the Pyramid, Revised and Updated 5th Anniversary Edition: Eradicating Poverty through Profits*, Upper Saddle River, NJ: Pearson Education.
Reficco, E. and P. Márquez (2012), 'Inclusive Networks for Building BOP Markets', *Business and Society*, **51** (3), 512–56.
Timmermans, S. and I. Tavory (2012), 'Theory construction in qualitative research: From grounded theory to abductive analysis', *Sociological Theory*, **30** (3), 167–86.
Van Maanen, J. (2011), 'Ethnography as work: Some rules of engagement', *Journal of Management Studies*, **48** (1), 218–34.
Vinck, D. (2012), 'Accessing material culture by following intermediary objects', in L. Naidoo (ed.), *An Ethnography of Global Landscapes and Corridors*, Rijeka, Croatia: InTech, pp. 89–108.

4. Staging co-design within healthcare: lessons from practice

Elizabeth B.-N. Sanders

OVERVIEW

My goal is to provide actionable insights for staging co-design with a focus on the front end of innovation in healthcare service design and delivery. These insights come from empirical evidence I have gained when staging co-design experiences for many different healthcare projects. I will share useful observations as well as principles for action. I will not address the literature on this topic in this chapter.

Three levels of engagement for staging co-design within healthcare are shown in Table 4.1. At the lowest level of engagement is the one-time event. This is typically a half-day, or a full-day, hands-on workshop with a small to medium sized group of people. The one-time event is often the first small step taken in approaching design from a participatory perspective for the purpose of innovation of products and/or services. The next level of engagement is a series of participatory events that take place over time and address different points along the design and development process. This level of engagement might include three or more meetings occurring every couple of weeks or months. The meetings could include the same people every time and/or new participants who join the later sessions. The highest level of engagement describes a co-design culture characterized by ongoing relationships between people. The co-design culture is markedly different from the other levels of engagement in that co-designing has become the default mode of engagement that is experienced by all who are involved.

The embodiments that emerge from the co-design process in healthcare differ across the three levels of engagement. I will describe these later and use photos from several projects in practice to show some of the embodiments created in the process.

The deliverables and potential impact at each level of engagement differ as well. The **deliverables** include both "hard" and "soft" deliverables. Hard deliverables could include ideas, product concepts, or service innovations.

Table 4.1 Three levels of engagement for staging co-design within healthcare

Level 1: one-time event	Level 2: iterative series of events	Level 3: co-design culture
Objectives: • Spark interest in the co-design process • Innovate products and/or services	Objectives: • Explore stages of the co-designing process • Innovate products and/or services	Objectives: • Seed and nurture the growth of a co-design culture throughout the organization • Innovate products, services and systems
Embodiments: • Dreams for future experience • Product and/or service ideas and concepts	Embodiments: • System and service journeys • Embodiments that change over the timeline of the co-design process	Embodiments: • Full-scale physical embodiments that support enactment • Virtual and augmented reality embodiments that support imagination and enactment • Emergence of a shared transdisciplinary language
Deliverables/impact: • Innovation in product and/or service design • Interdisciplinary collaboration • Identification of co-design players • Short-term impact	Deliverables/impact: • Innovation in product and/or service design and development • Collective creativity • Identification of dedicated team of co-designers • Planting seeds for a co-design culture • Ownership of the end results • Longer-term impact	Deliverables/impact: • Innovation in product, service and/or system design and development • Transdisciplinary collaboration • A cultural shift toward a co-designing mindset • Empowerment to enact change • Sustainable impact

Soft deliverables include empowerment (of the co-design participants) through engagement in the process as well as their ownership of the end results. **Impact** refers to what happens after the co-design engagement is over. Did anything change? Did anyone change? Does the project continue? If so, who takes charge and keeps the effort moving forward? The level of impact increases from the first to the third level of engagement. Since the type of impact is not known in advance, it is important to set up reasonable expectations from the start. For example, the emergence of a co-design culture will not occur after one series of iterative co-design events.

STAGING CO-DESIGN WITHIN HEALTHCARE

The basic components for staging co-design in healthcare include the actors, the place, the plan and the embodiments that emerge in the execution of the plan by the actors in the place.

The actors include all the people who are involved in the co-designing process. They fall into two categories: the participants and the facilitators (i.e., those who are responsible for staging the co-design process).

The participants include the people who will take part in the co-design process. In co-designing the participants include those who will be impacted by the results of the process. In healthcare design efforts, the participants might include patients (current and former), family members of the patients, physicians, nurses, social workers, nutritionists, cleaners, sitters, physical therapists, clergy, etc. The list is long. In a participatory design research project exploring the needs of all the people who use the hospital patient room, 25 different stakeholder groups were involved in full-scale co-design activities over a five-year period. The results and design implications have been published in a series of articles (Lavender at al. 2015, 2019; Patterson et al. 2017, 2018).

The participants differ in terms of the experience they bring to the co-design process. In healthcare design, some of them bring professional experience and others, such as the patients and family, bring lived experience. Both types of experience are equally important for relevant and sustainable innovation. The participants differ also in the attitude or mindset they bring to the co-design process. I have seen four distinct mindsets emerge with regard to the core concept of co-design, i.e., the idea that all people are creative and should be involved in the ideation, design and development of new products and services that will affect their futures. These four mindsets include intuitives, learners, sceptics and converts (Sanders and Stappers 2012).

The intuitives are fully aware that all people are creative. They don't need to be convinced that co-designing has value. In fact, they only need to be involved in one session before they try co-designing themselves. They may have been operating with a co-designing mindset all along without knowing there was a name for it. I have encountered more female than male intuitives in my practice.

The learners will come to understand and appreciate co-designing through several hands-on experiences. Their learning is accelerated if they are involved throughout the co-design process so that they can see the deliverables and impact of this approach.

The sceptics do not believe that all people are creative. It may be that they were trained to think of themselves as the experts in their domain and so they are not open to co-designing with the people they consider to be less knowledgeable. I have encountered many sceptics while practising co-design in the United States.

The converts are sceptics who, once they take part in the co-design experience, end up becoming extremely strong advocates for co-designing. But they may initially be disruptive of the process. It is difficult at first to distinguish the converts from the sceptics, so it is necessary to address everyone's questions and challenges. The converts can become very strong advocates for the co-design process and can be valuable actors in the higher levels of participatory engagement.

There is also fifth type of mindset regarding co-design. **The hypocrites** are those who pretend to believe, or who say they believe in the value of co-designing but don't really believe that it is of value. I have not observed this mindset in the United States. It was pointed out to me by a colleague working in Denmark.

The facilitators include the people who plan, conduct and document the co-design event(s). In other words, they stage the co-design process. The facilitators are likely to have different mindsets about co-designing. For example, we may have intuitives, learners, sceptics, converts, and/or hypocrites among the facilitators. It is important to recognize mindset differences and to strategically assign roles to the facilitation team members so that they can contribute effectively.

The most critical role is the **primary facilitator** because they are the ones who speak with the participants, giving them instructions and feedback as the session progresses. The primary facilitator *must* believe that all people are creative and that everyone has an important role to play in co-creation. A sceptical or hypocritical team member should not take on the primary facilitator role. The primary facilitator must also be a good listener and be able to adapt to the situations that arise spontaneously in co-design sessions.

The other roles needed on the facilitation team include recruiting, documenting, hosting, supporting, and time-keeping.

- **Recruiting** refers to the identification and selection of participants. It also includes whatever it takes to make sure that the participants arrive to the right place at the right time.
- **Documenting** refers to various ways that the events are recorded, such as note-taking, audio-recording, photography, video-recording, etc. Obtaining consent from the participants to record the activities is a requirement of this role.

- **Hosting** refers to making the participants feel welcome and comfortable. For example, it is always a good idea to offer the participants something to eat and drink as well as to make time for occasional breaks.
- **Supporting and time-keeping** refers to helping the primary facilitator keep the session on track and on time.

One of the first steps in staging the co-design process is to decide who will take on which role(s). In small co-design sessions, the facilitating team members may need to play more than one role but there should always be at least two members on a facilitation team, i.e., the primary facilitator and a versatile support person.

The place refers to the venue where the co-design event takes place. This includes the atmosphere of that environment and the furniture that is available for use. The place has a large impact on the success of the co-design session so it is important to visit the place ahead of time in order to be as prepared as possible.

There is not a set of specific guidelines that can be used to prepare the place since the actors, the plan, and the embodiments vary for each co-design event, but here are some questions to consider.

- Is the place big enough for the number of participants who are planning to attend? Is the place big enough for them to move freely when activities call for full body movement?
- Is the place handicapped accessible?
- Does the place provide natural lighting? Are there views and/or access to nature?
- Is the place flexible such that all the participants can gather at once yet they are also able to form smaller break-out groups?
- Are there enough chairs and tables for all?
- Is it OK to make a mess? Is it OK to make noise?
- Is it OK to pin or tape embodiments to the walls?

The plan includes all the activities needed to get ready for the event, the facilitation of the event, the documentation of what takes place during the event and the analysis of the outcome. The plan also includes the procurement of and any preparation of the materials that will be used for the embodiment of ideas as well as the script that the facilitator will use in guiding the session. The plan should be written in detail before each event so that everyone on the facilitation team knows what will happen, where it will take place, how long it will take and who will be involved. The plan also needs to stay open to the unexpected. Experienced facilitators are needed for addressing the unexpected.

The plan varies according to the objectives of the event. For example, it could be focused on exploring the future of the patient experience in an outpatient clinic. Or it could be about the transition to new work practices of staff in a unit within the hospital. And at a large level of scale, the plan could describe the design of an entire healthcare hospital campus.

There is a lot more to discuss in relation to planning co-design events than these few observations. The interested reader will find a wealth of suggestions and ideas for planning in *The World Café: Shaping our Futures through Conversations that Matter* (Brown et al. 2005) and *Convivial Toolbox: Generative Research for the Front End of Design* (Sanders and Stappers 2012).

Embodiments refer to the “things” that are made by the participants in the co-design process. I will use the word “embodiment” instead of “object” because it calls to mind a wider variety of things that might be made.

> The embodiment of something gives concrete form to an abstract idea. A flag is the embodiment of a country. When you talk about embodiment, you’re talking about giving a form to ideas that are usually not physical: like love, hate, fear, justice, etc. A gavel is the embodiment of justice; a wedding ring can be the embodiment of love. The word body in embodiment is a clue to its meaning: this is a word for giving a body to things that usually don't have one. (Vocabulary.com n.d.)

Embodiments that play out in the front end of the co-design process take on many different forms throughout the process including visualizations of experience (e.g., dreams and fears), mappings of relationships between people, rough ideas for products, sketchy concepts for frameworks, early prototypes and subsequent solutions. These types of embodiment are created by the co-design participants using generative materials that have been carefully staged to support, facilitate and catalyse their collective creativity. Embodiments that are made from generative materials give presence to thoughts and feelings and make it easier for people to share ideas with each other. This is especially important in interdisciplinary collaboration where language may be a barrier.

Embodiments in the front end of the co-design process will emerge from the participatory prototyping cycle (PPC) of making, telling and enacting (Brandt et al. 2012). The PPC is a staging framework that invites the relevant actors into the co-design process, and provides them with generative materials that they can use to embody their ideas. The PPC combines making, telling and enacting and uses each activity to fuel the next. The participants do not need to have experience as designers to embody their ideas. By putting making together with telling and enacting, people who are not skilled in making are able to externalize their ideas and to do so in collaboration with others. The primary advantage of the making/telling/enacting cycle is that it provides alternative forms of expression for actors in the co-design process. Some people respond best to stories, some to enactments and others to physical embodiments (Figure 4.1). By utilizing all three in an iterative cycle, everyone who has a stake in the future can contribute to the generation of ideas.

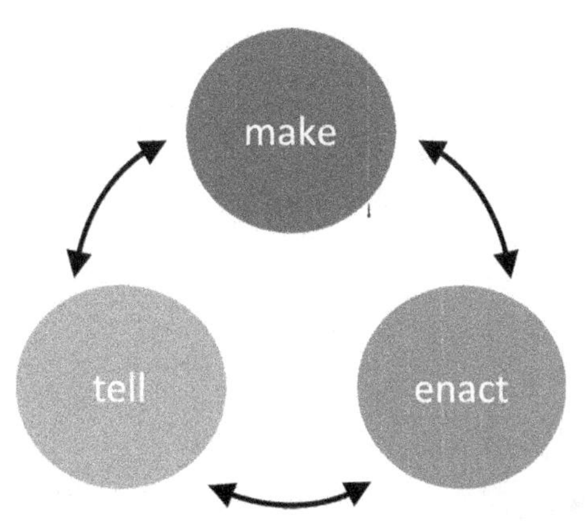

Note: The PPC helps people to embody their thoughts, feelings and ideas for current and future experience.

Figure 4.1 The participatory prototyping cycle (PPC)

Examples of some of the embodiments that I have used in healthcare design projects are shown below and described in the photo captions (Figures 4.2 to 4.5).

LEVEL 1: STAGING A ONE-TIME CO-DESIGN EVENT

The one-time co-design event is the lowest level of participatory engagement. It usually takes place early in the design process where participants are brought together to collaboratively imagine and embody their dreams for future experience. It may take place over three to four hours but a better session length would be a full day. If the participants have been prepared for the event, for example, though "homework" that asks them to reflect on their past and current experiences relating to the topic at hand, the timing of the event can be shortened. It is a good idea to begin to explore the co-design approach with a one-time event since it is lower in cost and risk than larger efforts. It is an excellent way to see how an organization might adapt to a co-design process vs. the more traditional design process.

The principles (shown in bold type) for staging a one-time co-design event are described below. They are organized by the key components: actors, place, plan and embodiments.

Note: A team of participants representing all areas of the hospital worked collaboratively to visualize the experiences that they wanted patients and family members to have when the new hospital was built, which was to be about eight years after this co-design workshop took place. In the photo, one of the participants presents the team's work by telling what the images and words mean to them at each level of the bulls-eye target. This collective embodiment inspired the design of the new hospital which has since been built.

Figure 4.2 Making and telling

The Actors

Clearly define and agree upon the role (or role combination) for each member of the facilitation team

The primary facilitator must believe that all people are creative and that everyone has an important role to play in the co-design process. Be sure that someone is assigned to recruit, facilitate, document, host, support and keep the time. Agreeing on the roles beforehand will help to avoid awkward moments and wasted time.

Recruit the participants at least three weeks in advance so that you have time to send them homework that will prepare them for being creative and expressive about the topic

The homework should ask them to reflect on their current situation and the past events that may have had an impact on their thoughts and feelings about the situation. Another goal of the homework is to get them curious about what will happen in the event.

Note: Healthcare providers who work every day in the Intensive Care Units (ICUs) at The Ohio State University Wexner Medical Center embodied their ideas about the flow of materials in the ICU room using simple 3D toolkit components. They first explained to the staging team how they currently handle trash and waste in the ICU room by drawing the purple trash flow lines on the "floor". They then explored many different ideas for improving this process in the future by using the 3D components and the staff and patient dolls to make and enact their ideas.

Figure 4.3 Making and enacting

Anticipate that your participants will come with different mindsets about co-design

Provide support for all types of mindsets. Manage their expectations by letting them know what will take place before, during and after the event. The homework will help everyone be able to participate successfully even if they come to the session with a sceptical mindset.

The Place

The one-time co-design event should be held at a "special" place that is outside the day-to-day working environment of the participants

The atmosphere and setting of the place should be comfortable. Fresh and healthy food and drink should be available for all. Access to and views of nature should also be available.

There should be spaces for sitting, working, playing and moving

Making can be messy so you will need to provide adequate space that is easily cleaned. Enacting takes open space and props for role playing. Encourage movement by providing a big, open space.

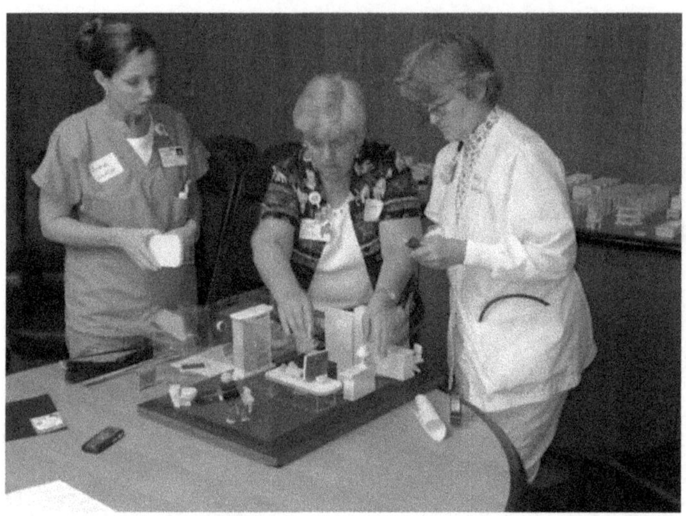

Note: Nurses who worked on different patient floors of a heart hospital were invited to collaboratively create an ideal patient room for the future hospital. The 3D toolkit contained both concrete and abstract room components to help bridge the gap between the current situation and future possibilities. The nurses were able to collaboratively create their future patient room in about ten minutes. It took them as long to present their room concept (by telling and enacting) as it did to make it.

Figure 4.4 Making, telling and enacting

The Plan

Plan the event with the place(s) in mind

Visit the place beforehand so that there are no surprises on the day of the event. Be sure that the participants have good directions on where to go and when to arrive.

Plan the activities in the co-design workshop with the participants' abilities and needs at the core

If your participants do not already know each other, plan one or two informal activities at the start of the session for them to get to know one another in meaningful ways.

Start the session with individual activities. Then ask the participants to share the embodiments of their individual activities with one another. Work toward collaborative activities once they have gotten the chance to hear from each other. At the end of the workshop provide an activity that brings all the teams' work together.

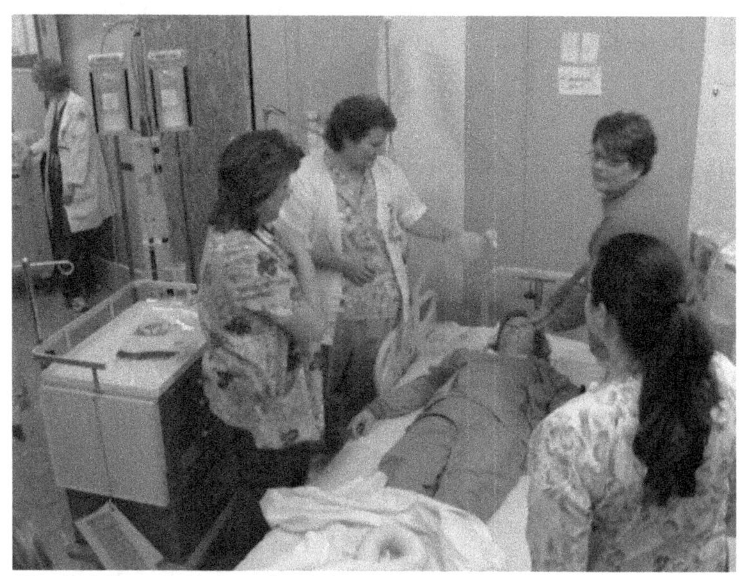

Note: The participants in this future hospital room scenario are all nurses who played the roles of the patient, physician and nurses. They were doing so to evaluate the proposed design and layout of the room, which had been embodied in rough, full-scale form. They enacted the key scenarios that would take place in the room in the future to make sure that the room "worked" before the architectural team proceeded to the next phase in the design process.

Figure 4.5 Enacting and telling

Put the agenda for the session in writing and review it with the entire facilitation team. The plan should include scripts for instructing the participants and time estimates for every activity.

Use the PPC of making, telling and enacting to structure the plan
Invite the participants to make things to express their thoughts, feelings and dreams. Allow the participants to explore their current and past experiences before asking them to explore future experience. Invite them to tell stories about the past, the present and the future. Let them enact past, present and future scenarios. Do not make enactment a requirement for everyone as it may be difficult for introverts to take part in activities such as improvisation.

Have backup activities and materials on hand in case the agenda needs to change
The timing of the session is often the hardest part to learn. It is a good idea to have extra activities prepared in case the plan goes more quickly than

expected. However, this is rare. And it is also wise to prioritize the various activities in case something needs to be dropped from the agenda in order to stay on schedule.

Document everything!
Use video-recording, audio-recording and photography. If one medium fails, you will still have documentation of the event. You do not need to analyse everything that is recorded but you will have the option to do so at a later point in time.

Pilot test!
Pilot testing of the plan and the generative materials can take place with people other than the actual participants. Revise the plan, the materials and the place, as needed, based on feedback from the pilot testing.

Generative Materials to Support Embodiment

Create custom-made toolkits from generative materials for each activity in the session. For more information on this, the reader is referred to *Convivial Toolbox* (Sanders and Stappers 2012).

Deliverables/Impact

The deliverables of a one-time co-design event will depend on the objectives set forth in the plan. In general, you can expect to see innovation in the front end of the design process that is achieved through interdisciplinary collaboration. For example, the embodiments may include collective visualizations of future dreams that can serve to inspire the process as it moves forward. Such visualizations may take the form of landscapes, roadmaps or journeys. The impact of a one-time co-design event is limited but many ideas may emerge. In addition, you may be able to begin to identify the intuitives, the learners and the converts among the participants. And you might be able to see evidence that you were able to plant the seeds of co-design in some of them.

LEVEL 2: STAGING ITERATIVE CO-DESIGN EVENTS AND ACTIVITIES

The second level of engagement is a series of participatory events that take place along the co-design process. There is usually a core group of dedicated participants who are committed to this effort as well as commitment from the organization in which the process is taking place to keep the co-design process ongoing. The staging of iterative co-design events must be adaptive to changes in the ongoing design process. The principles for staging iterative co-design events are described below.

The Actors

Identify and celebrate the intuitives and the potential converts

The intuitives and the potential converts are likely to be those who catalyse the co-design approach and support it internally. They may be the early adopters of the co-design culture who will work behind the scenes and in between the meetings to keep the effort moving forward. So it is important to recognize and reward their contributions. At the same time, it is important to address the questions from the sceptics in order to nurture those with the potential to become converts.

Invite new perspectives to the core team of actors

It is ideal to have some people who take part in every event to provide continuity. But at the same time you will want to invite newcomers to the process. They can add fresh ideas and constructive feedback regarding what has already taken place.

The Place

Have a dedicated place where the actors can engage with each other whenever they choose to do so

On-line communication is good for updates and previews, but face-to-face meetings are best for collaborative work. Having a dedicated place also gives the actors the opportunity to make, tell and enact together whenever the opportunity arises.

The dedicated space can then become the repository for all the embodiments, including intermediary objects such as summary materials, that emerge from the co-design process. The embodiments will provide inspiration to spark the creativity of the actors. The embodiments can also serve as a learning environment for the new team members to immerse themselves in the project and the co-design process.

The design of the place becomes even more important when you are staging iterative co-design events and activities

With a large, flexible and dedicated space, you will have the opportunity to design the space to simultaneously support various modes of work. There are excellent sources available now to help you design such a space. One that is particularly useful is Make Space (Donley and Witthoff 2012). You might also draw inspiration from the spaces, places and materials in Reggio Emilia inspired schools (e.g., Caldwell 1997). Many such inspirational photos have been posted on Pinterest (e.g., see Reggio Emilia classroom).

The Plan

The plan should address the entire time course of participation

You will need to plan not only the co-design events but also what happens in between the events. Before every event there should be an activity such as homework to stir curiosity and excitement about the upcoming event. After each event there should be a summary (such as a report, a series of posters, a video, etc.) that the participants can share with others. Incidentally, sharing summaries can be a good way to recruit more participants who are interested in engaging in future co-design events. In between events, the dedicated space (that is by now full of embodiments of the process) should be available for spontaneous collaboration sessions and for sharing the co-design process with other interested people.

Generative Materials to Support Embodiment

Provide a toolbox of materials from which the actors can construct custom-made toolkits for the purpose at hand

For a complete list, including photos, of the generative materials that are good to include in a co-design toolbox, the reader can refer to Sanders (2018).

The embodiments of the co-design process change over the time course of participation

In the beginning the embodiments will be of abstract experience and they will become more concrete over time. The forms of making used most often early in the design process include maps, timelines and collages. Later forms of making include props, Velcro-models and really rough prototypes. The traditional forms of prototyping such as sketching and model-making appear later in the co-design development process. In the front end of the co-design process, the types of telling come in the form of stories. Later forms of telling might include descriptions of the imagined products, spaces, etc. Even later forms of telling include presentations and selling events. Early in the co-design process, the forms of enactment can be best described as “pretending”. Later forms of enactment might include improvisation and performance.

Deliverables/Impact

The deliverables of an iterative co-design process may include any or all of the following: hard deliverables such as new products, services, processes and systems, as well as soft deliverables such as empowerment of those involved and ownership of what is being designed. In the early front end of innovation, the most useful deliverables of the iterative co-design process are often guidelines and principles for design. The impact of an iterative co-design process might include the identification of the lead proponents and the emergence of seeds of a co-design culture.

LEVEL 3: LAYING THE GROUNDWORK FOR A CO-DESIGN CULTURE

The highest level of engagement is a co-design culture characterized by ongoing participatory relationships. This is a new culture, one that is founded on the shared belief that all people are creative and should be involved in design for the future.

The Actors

The lead actors in the emerging co-design culture will self-identify and it is likely that they will be the intuitives, the learners and the converts. We will need to support them so that they can continue to grow as the co-design culture emerges. It will take time for this transformation occur.

The Place

As the culture shifts to a co-creative worldview, we will need to address the places where co-designing happens. For example, the architecture of healthcare environments needs to provide for a variety of spaces where people can engage with one another through making, telling and enacting. These places should allow for full-scale prototyping of proposed healthcare environments at the earliest stages of the process. There should be spaces where people can engage collectively and express themselves with whole body movement as they enact future scenarios. Generative materials for making need to be always available in the co-design spaces and places. The places where co-designing happens must address people's wellbeing and support restorative functioning through biophilic design considerations.

The Plan

A culture of co-design will become a way of life so there is less need to make explicit plans to bring people together to make it happen. Collaboration will become the default mode of interaction. The co-designers will need feedback on how they are doing and how the cultural transformation is going. We will need to assess the transformation and to identify and reward the cultural change agents.

Generative Materials to Support Embodiment

Since my practice has taken place mainly in the United States, Level 3 remains an aspirational goal since co-designing in healthcare is just now beginning to take hold. So I have no photos of embodiments to share. Co-design cultures within healthcare have made more headway in other parts of the world such as in Norway (see www.ccsdi.no) and New Zealand (Reay at al. 2017). It is my hope that the countries who are leading this transformation will continue to share what they are doing and how it is working.

In the future, the embodiments that arise from the co-design culture will be made from a much wider array of materials. For example, augmented reality, virtual reality, artificial intelligence and mixed reality will provide opportunities for digital embodiments of people's hopes, dreams and fears. We will explore the new digital materials for embodiment and investigate their potential for supporting collective visualizations. The new technologies will enable us to experience full-scale spaces and places where we can enact future scenarios before they actually happen. These embodiments will change over time as new technologies become available and as new actors join the co-design culture. The full-scale embodiments (both physical and virtual) will provide the stage upon which everyone can be an actor in the future of healthcare.

Deliverables/Impact

A co-design culture within healthcare has the potential to positively impact all the actors in this context including patients, family members and healthcare providers. The generative materials will provide the basis for a transdisciplinary design language that can be used by everyone. New healthcare products, environments and services will simultaneously address the needs and dreams of all the people. Having a co-design culture within healthcare will give the co-designers ownership of what they have made together and it will empower them to make changes that are needed as the future continues to unfold and emerge.

CONCLUSIONS

This chapter provides insights and principles, derived from practical experience, for those who are interested in staging co-design within healthcare. It points to a path for learning in which co-designing starts with one-time co-design events before moving to an iterative co-design process and ultimately aims toward the seeding and emergence of a co-design culture. Table 4.1 provides a high-level summary of the objectives, embodiments, deliverables and impact that differentiate the three levels of engagement for staging co-design within healthcare.

REFERENCES

Brandt, E., T. Binder and E.B.-N. Sanders (2012), 'Tools and techniques: Ways to engage telling, making and enacting', in J. Simonsen and T. Robertson (eds), *Routledge International Handbook of Participatory Design*, London: Routledge, pp. 145–81.

Brown, J. with D. Isaacs and the World Café Community (2005), *The World Café: Shaping our Futures through Conversations that Matter*, San Francisco: Berrett-Koehler Publishers.

Caldwell, L.B. (1997), *Bringing Reggio Emilia Home: An Innovative Approach to Early Childhood Education*, New York: Teachers College Press.

Donley, S. and S. Witthoff (2012), *Make Space: How to Set the Stage for Creative Collaboration*, New Jersey: John Wiley & Sons.

Lavender, S.A., C.M. Sommerich, E.S. Patterson, E.B.-N. Sanders, K.D. Evans, S. Park, R.Z. Radin Umar, and J. Li (2015), 'Hospital patient room design: The issues facing 23 occupational groups who work in medical/surgical patient rooms', *Health Environments Research & Design Journal*, **8** (4), 98–114.

Lavender, S.A., C.M. Sommerich, E.B.-N. Sanders, K.D. Evans, J. Li, R.Z. Radin Umar, and E.S. Patterson (2019), 'Developing evidence-based design guidelines for med/surg hospital patient rooms that meet the needs of staff, patients, and visitors', *Health Environments Research & Design Journal*, **13** (1), 145–178.

Patterson, E.S., E.B.-N. Sanders, C.M. Sommerich, S.A. Lavender, J. Li, and K.D. Evans (2017), 'Meeting patient expectations during hospitalization: A grounded theoretical analysis of patient-centered room elements', *Health Environments Research & Design Journal*, **10** (5), 95–110.

Patterson, E.S., E.B.-N. Sanders, S.A. Lavender, C.M. Sommerich, S. Park, J. Li, and K.D. Evans (2018), 'A grounded theoretical analysis of room elements desired by family members and visitors of hospitalized patients: Implications for the medical/surgical hospital patient room', *Health Environments Research & Design Journal*, **12** (1), 124–144.

Reay, S., G. Collier, J. Kennedy-Good, A. Old, D. Reid, and A. Bill (2017), 'Designing the future of healthcare together: Prototyping a hospital co-design space', *CoDesign: International Journal of CoCreation in Design and the Arts*, **13** (4), 227–44.

Sanders, E. (2018), 'A generative toolbox for social innovation', *Current 08*, accessed 1 November 2019 at http://current.ecuad.ca/author/elizabeth-sanders.

Sanders, E.B.-N. and P.J. Stappers (2012), *Convivial Toolbox: Generative Research for the Front End of Design*, Amsterdam: BIS Publishers.

Vocabulary.com (n.d.), 'Embodiment', accessed 1 November 2019 at https://www.vocabulary.com/.dictionary/embodiment.

5. Circulating objects between frontstage and backstage: collectively identifying concerns and framing solution spaces

Signe Pedersen and Søsser Brodersen

INTRODUCTION

Co-design in a healthcare setting is often very complex, as many disciplines, backgrounds and competences are at play. Participatory design practitioners advocate that future users as well as other actors should be involved throughout the design process, so that the new concept can be shaped from multiple perspectives (Bratteteig and Gregory 2001). However, multiple and diverse concerns shape not only the concept or solution, but also the 'problem' that the concept is intended to address. The design thinking literature highlights that problems are defined based on initial empirical investigations; likewise, scholars from the engineering design tradition have illustrated how problems and solutions co-evolve over the course of a design project (Dorst and Cross 2001). Thus, when engaging multiple and diverse actors in participatory design processes, the complexity in terms of understanding problem(s) as well as solutions increases. In this chapter, we draw upon actor–network theory (ANT) and the notion of intermediary objects (Boujut and Blanco 2003) to develop a conceptual framework of frontstage and backstage activities in design to investigate how problems (matters of concern) and solutions are negotiated through a process of producing and circulating objects. We follow and investigate the efforts of five engineering design students striving to collaboratively design sensory stimulation technologies in a nursing home setting as an exemplary case to unfold the framework.

OBJECTS IN DESIGN

Materiality and objects have always played essential roles in design activities, and prototypes and other objects such as design games (Brandt et al. 2008) can be viewed as key actors in many participatory design projects. Objects can help facilitate dialogue especially in healthcare settings, which involve vulnerable actors. For example, in nursing homes, patients and elderly people with dementia might struggle to express their viewpoints due to cognitive impairments. Design games and prototyping have proven to be useful for engaging vulnerable actors, as well as for facilitating mutual learning and negotiating concerns (Pedersen 2020). We conceptualize activities whereby objects are circulated between actors during design events such as ethnographic field studies or workshops as frontstage activities, because this is where design efforts become visible to non-designers and begin to perform in a collaborative way. In the case presented in this chapter, these non-designers are different actors affiliated with a nursing home in Denmark.

In engineering design, prototyping and other design methods such as affinity diagrams, sequence models, and morphology charts produce objects which are displayed in design studios to promote knowledge sharing in design teams. We refer to such activities as backstage activities because some of the design efforts might unfold 'behind the scenes'—that is, out of the immediate sight of future users of the solution. In line with frontstage activities, these backstage activities take place in heterogeneous networks and thus involve negotiation between humans and non-human actors such as design objects (Latour 1992).

In many engineering design projects, objects such as prototypes are produced backstage to be circulated frontstage (e.g., in test situations with future users). In participatory design, for instance, design games are produced backstage to be played and thus translated through engagement with users and other actors. In both instances, the objects are made to perform in front of or in collaboration with an audience, often consisting of potential users. In this chapter, we argue that when numerous actors are involved in collaborative design work, the exploration of problems (concerns) and solutions can be understood as an iterative process of continuously shifting between frontstage and backstage activities. We illustrate this point by drawing upon the theatrical metaphor in line with the ideas presented by Pedersen, Dorland and Clausen in Chapter 2 (this volume) and investigate how objects are staged in terms of being inscribed with concerns, insights, and potential solutions, and subsequently circulated to perform as intermediary objects in negotiations amongst a range of important actors from the healthcare context.

The analytical framework of intermediary objects (Boujut and Blanco 2003; Vinck 2012) draws upon ANT (Latour 1987) to analyse design situations by following objects as they are circulated between actors in and across networks. Objects perform as intermediary objects when they mediate and foster knowledge-building in the design process. They have three main features: (a) they can mediate between actors by providing a shared reference point (e.g., for designers and users); (b) they can translate actors' positions (intentionally or unintentionally); and (c) they can represent ideas as well as processes

(Vinck 2012). In this chapter, we use the analytical concept of intermediary objects to analyse the design efforts of a team of five third-semester engineering design students as they staged and facilitated negotiations of matters of concern among nursing staff, elderly residents with dementia, and managers in a Danish nursing home. The case further illustrates how solutions are negotiated through a process of producing and circulating objects.

METHODOLOGY

One of the authors supervised the five design students from the engineering education programme in a sustainable design course at Aalborg University. Throughout the 12-week semester, the students were introduced to theory about user involvement, participatory design methods, and prototyping techniques. They had no prior knowledge about these subjects before the semester began. Our analyses and arguments are based on the students' written work (Olsen et al. 2016), observations made by one of the authors while the students implemented their design project, and qualitative interviews with nursing home staff.

CASE STUDY: ATTEMPTING TO IMPROVE QUALITY OF LIFE FOR ELDERLY RESIDENTS WITH DEMENTIA

Approximately 75–90 per cent of residents in Danish nursing homes have been diagnosed with dementia. Dementia causes severe challenges for those who have been diagnosed, their relatives, and nursing staff, as the condition impedes their ability to express certain needs, wishes, and inconveniences. Often, this inability to communicate leads to feelings of restlessness, frustration and anger (Alzheimer's Association 2020). Experiences show that one way to ease these feelings of restlessness and frustration among elderly residents with dementia is to use sensory stimulation technologies, such as the Paro seal, music, and light sensors. The nursing home management had bought a number of these promising technologies and placed them in a special room in the nursing home dedicated to such technologies; however, they felt that they needed more knowledge about how to use them and when.

Thus, the concern of how the residents and staff could benefit from using these sensory stimulation technologies served as the starting point for a collaboration between the student design team and the nursing home. Together, the students and the nursing home management discussed and negotiated a common matter of concern: How can sensory stimulation technologies be used to ease feelings of restlessness and frustration among elderly residents with dementia?

Initial Fieldwork

The first visible concern that the students identified during their initial research at the nursing home was that although the nursing home had sensory stimulation technologies onsite, in reality the technologies essentially remained unused and thus did not provide any ease for the elderly residents with dementia. The students documented their findings as they took photographic field notes documenting the room filled with technologies—but no nursing home residents in sight—and wrote thick descriptions of what they observed. Thus, the students produced objects onsite in the form of pictures and narratives that helped them capture a visible concern at the nursing home. Back at their design studio, the students printed the pictures, synthesized their thick descriptions into cohesive narratives and added these insights to a physical affinity diagram that hung on a wall in their design studio. They had already created this diagram based on their initial desk research and now began negotiating this knowledge with their empirical data. They structured the affinity diagram based on themes that had emerged from the different physical bits and pieces comprising their empirical material. The affinity diagram thus became an intermediary object consisting of ordered representations (in the form of pictures and narratives) of the situation at the nursing home. In this situation, the affinity diagram mediated the insights gathered between the members of the design team and translated their observations into categories of empirical material while evoking alignment among members of the design group.

Staging Negotiations with Elderly Residents with Dementia

Backstage: creating a game as an intermediary object

The students interpreted the result of their negotiations with each other and with the affinity diagram as follows: one reason for this lack of use might have been the challenge of getting the residents out of their apartments and into the room. The students wanted to test this assumption and decided to stage negotiations with nursing home residents. They framed these negotiations as providing opportunities for residents to describe their experiences with the technologies and to express what made them feel happy and comfortable in their daily lives.

The students knew that it would be difficult to stage these negotiations with the nursing home residents. Engaging elderly people with dementia is challenging because the disease makes it difficult to interact through typical conversation (Hendriks et al. 2014, 2015; Lindsay et al. 2012; Robinson et al. 2009; Rodgers 2018). It was nevertheless important for the team to provide residents with opportunities to voice their concerns. Thus, they began to develop objects in the form of a design game which they inscribed with meanings based on their knowledge of how to communicate with residents with dementia. The students created 'game pieces' consisting of photos of different technologies and activities, and hoped that when combined with a 'ladder' diagram intended to frame and prioritize residents' interests and concerns, the game would serve as an effective intermediary object. The students hoped that the game would represent what they imagined would be fun activities for the residents, such as walking, dancing, playing board games and pool, interacting with the Paro seal, etc., thereby mediating between themselves and the residents.

Frontstage and backstage: testing and reframing the design game

The students invited the nursing home residents to 'play' their game in the common room, where the TV was on and other residents and staff were seated and walking around, making the room a bit noisy and crowded. They placed the game on the table and asked the residents to rank the activities which they enjoyed the most. However, due to the abstract nature of the activities and the 'ladder' ranking approach, the complexity of the game exceeded residents' capacity. Residents were easily distracted due to the chaotic nature of the surroundings, and simply did not know what to do. Others just stared at the design team. Thus, the design game failed to perform as a meaningful and successful intermediary object, as it did not mediate between students and residents, meaning that the residents' concerns were neither articulated nor translated to the design team.

The students left the nursing home and reflected on their unsuccessful performance. They used their insights to reframe the design game by adding smiley face and frowny face icons for the residents to indicate whether they liked the activities illustrated in the photos. They set up a new engagement with the nursing home and tried the new game, but it was still too complex, and the common room was a distraction once again. Back at the design studio, they reframed the design game a second time by inscribing less complex and abstract meanings into the game, and made arrangements with the nursing home to play it in the calm environment of the residents' private living spaces.

In this particular setting, the components of the reframed design game finally worked as intermediary objects that enabled the residents to express what made their lives meaningful and what would make them feel at ease. The residents selected the game pieces representing the activities they enjoyed the most, including social activities (e.g., excursions, playing games, walking), being independent (e.g., dressing and bathing themselves), and being understood (i.e., communication). Thus, the students eventually managed to navigate the situation by staging and facilitating negotiations and reframing the game based on lessons learned from previous efforts. Interestingly, although sensory stimulation technologies were a concern for nursing home managers, none of the residents associated these technologies with activities that had been identified as key matters of concern.

Backstage: synthesizing and reframing concerns

Having identified some of the key matters of concern among elderly residents with dementia, the design team began to update their affinity diagram with the new insights in the form of new pictures and quotes from the nursing home residents. Through this process, the affinity diagram came to represent the concerns of both residents and managers of the nursing home, as well as desk research about dementia and sensory stimulation technologies. Yet again, the field notes and pictures generated as a result of frontstage negotiations with actors (i.e., nursing home residents) represented the concerns of the actors engaged in co-design activities, and mediated between the members of design team who had facilitated the frontstage negotiations and the team members who had not participated. Furthermore, post-it notes representing the insights mediated and translated between the students and their affinity diagram, making it a process of collective interpretation and reframing.

Negotiating with the affinity diagram by rearranging the post-it notes representing the empirical data led to a reframing of the design focus, as desk research combined with residents' concerns about independence and communication revealed possible tensions among nursing home residents and nursing staff. In some cases, frustrated residents yelled at or even hit the nursing staff members who were trying to assist them. Thus, the backstage activities of synthesizing knowledge in an affinity diagram led members of the design team to reframe their focus toward improving the interactions between the nursing staff and elderly residents to *avoid potential conflict, thereby easing tensions and improving residents' quality of life.*

Understanding the Relations between Residents and Nursing Staff

The next step for the design team was to engage with another set of key actors—the nursing staff. Staff members play a central role in the lives of the residents and they often come to know the residents very well, making them well-positioned to potentially speak on behalf of elderly people with dementia. Furthermore, members of the nursing staff often support the residents by helping them engage in activities that are important to them (i.e., social activities, being independent, and being understood). However, they can also directly or indirectly hinder the residents from doing what they want, either by not having time to engage in certain activities at a given moment, or by failing to understand what residents are trying to express. Thus, the design team wanted to further explore the relationship between the residents and the nursing staff, and decided to contact them directly to learn about their concerns.

Frontstage: observing the activities of nursing staff

The design students made arrangements with the nursing home management and nursing staff to spend more time at the facility in order to gain a better understanding of residents' everyday lives, with a particular interest in the interactions between residents and nursing staff. The students followed staff members for one week as they performed their daily routines. They focused particularly on situations where they sensed residents becoming agitated and frustrated, as these would be the situations where sensory stimulation technology might be helpful. By following the nursing staff and residents, it became apparent to the design team that three types of situations elicited feelings of frustration from residents: (a) engaging in the morning routines of waking up and getting dressed, (b) needing to leave the nursing home (e.g., for a doctor's appointment), and (c) spending time in the activity centre after breakfast. In these specific situations, the team observed some residents expressing feelings of anger and frustration to the extent that they attempted to hit the nursing staff or throw objects.

Backstage: renegotiating with the affinity diagram

After returning to the design studio and updating their affinity diagram with new pictures, quotes, and thick descriptions, the students began staging a new interaction with the staff. They *interpreted* their findings from the fieldwork by coming up with the hypothesis that residents' discomfort and frustration might be due to them not being understood in certain situations. In such situations, their anger and frustration made them unwilling to participate in any activity suggested to them, including activities involving sensory technologies.

The design team thus *framed* the negotiations by *inscribing* a design game inspired by a sequence model. This object was intended to encourage nursing staff to speak freely about conflicts and concerns related to three situations: engaging in morning routines, needing to leave the nursing home, and spending time in the activity centre. The design team divided the three situations into sub-tasks and wrote them on game pieces (e.g., 'changing diaper', 'lifting from bed to wheelchair', 'making the bed', 'brushing teeth'; 'washing face', etc.). They also made smaller pieces to represent challenges (visualized as 'bumps on the road'). The game was scripted to frame a dialogue about daily routines, thereby providing a way for nursing staff to indicate situations where challenges typically arose.

Frontstage: using the design game as an intermediary object

The nursing staff easily understood the design game and began to move around the game pieces and discuss situations that were particularly challenging and factors contributing to those challenges. Thus, the design game was 'circulated' in the network of nursing home staff and served as an intermediary object that represented challenges that the staff could recognize, and mediated between the staff members as well as between the nursing staff and the designers.

The staff explained that in these three situations, residents have very little control, as they are 100 per cent dependent on the nursing staff, which triggers the concern of independence expressed by the residents. It also became apparent that the nursing staff often felt conflicts arise in one-on-one situations, making it difficult for them to engage in meaningful care. Interestingly, during the workshop, the nursing staff reframed the discussion from challenges to potential solutions. They suggested the idea of a mobile sensory stimulation technology that could be used in the residents' apartments where many of the conflicts arose.

The team immediately followed up on this idea and abandoned the design game while initiating a dialogue about a potential future solution, including functions and uses. They placed ideas on post-it notes, making them visible to all actors in the room, and solicited comments and feedback. In this situation, the nursing staff and the design students produced new intermediary objects in-situ and inscribed the concerns of the nursing staff into them. With this new development, the design team headed back to their design studio.

Developing a Prototype

Backstage: producing a design specification

By the time the students began to shift their attention from staging negotiations about problems or concerns towards imagining potential solutions to these concerns, another central object was produced by the design team: a design specification. The students collectively interpreted the insights from the affinity diagram and translated these into a design specification. This design specification served as a central intermediary object that translated the insights from the affinity diagram into a manageable and forward-pointing representation of the synthesized empirical data from the engagements and field studies at the nursing home. During this process, the insights were reframed as requirements and desired characteristics of the future solution. Thus, this design specification also played a role in framing future concepts generated during brainstorming sessions, negotiating these concepts, and selecting the final solution. As was the case with the affinity diagram, the design specification was not a static object that was inscribed only once. Rather, it was a dynamic object that was continuously negotiated based on new insights and ideas. It thus comprised the results/reframing of frontstage as well as backstage negotiations.

Backstage and frontstage: producing and testing prototypes

During a brainstorming session, the design students translated the requirements and desired characteristics from the specification into three different concepts, each focused on a different sensory stimulus: lights, sounds, and tactile sensations. They produced rapid prototypes to represent each concept and staged a meeting with nursing staff to select a specific staff member to serve as the 'prototype ambassador' for each one. These three prototype ambassadors were responsible for testing the prototypes and engaging residents and fellow staff members in the process. This was a smart strategic move, as the students managed to stage this intervention by engaging the staff in circulating the prototypes, thereby giving them ownership and a sense of worth in the design process. The team did not interfere in the testing phase; in fact, they were not present at the nursing home, leaving the circulation of the objects solely to the appointed ambassadors.

After the 2-week testing period, the team staged and facilitated follow-up dialogue with the nursing staff and learned that the prototypes had worked very well as intermediary objects between designers, staff, and residents. The prototypes managed to represent diverse design ideas, and played a mediating and translating role, especially for the nursing staff, who were eager to share insights from this small test with the students. Their feedback revealed that all three sensory components (i.e., lights, sounds, and tactile sensations) were important aspects to incorporate in the final design.

Backstage and Frontstage: Ideation

The students immediately incorporated this feedback directly into the design specification, making it more detailed and precise as they incorporated additional requirements and desired characteristics. The outcome of these negotiations based on the first prototype test directly translated the design specification as well as the concerns and motivations of the students. As the design specification was updated with more empirical data about the network where the solution needed to fit, the solution space narrowed. Thus, the design specification played a central role in influencing additional staging efforts in the design process. The students translated the insights and concerns represented by both the affinity diagram and the design specification into a morphology chart with pictograms representing six themes from the design specification: mobility, functionality, situations, shape, instructions, and flexibility. This morphology chart was meant to function as an intermediary object to be circulated in frontstage activities spanning a small ideation workshop with two nursing staff members and two musical therapists.

During the session, many interesting ideas were generated. However, most of the solutions suggested by the staff were based on existing solutions. Thus, the morphology model proved too narrow and specific, and framed the negotiations to the extent that the participants were not encouraged to think outside the box. Back in the design studio, the students decided to elaborate on the ideas generated by the nursing staff and eventually came up with three concepts. The students ensured that the ideas generated by the staff were inscribed in—and thus also represented by—the concepts, so the staff would recognize their inputs and feel that the concepts represented their concerns.

The design team then developed three concepts that each had their own characteristics and properties: a lamp-inspired stimulation; a radio-inspired stimulation, and a textured device shaped like a stone. The students brought sketches of the three concepts to a mini-workshop with representatives from the nursing home staff. During this session, the nursing staff very clearly preferred the textured stone concept. They liked it because it provided two types of stimulation (i.e., vibration and light) and was highly portable, and thus could be used in different settings. However, it needed to be lightweight, in case residents decided to throw it at the nursing staff.

Backstage and Frontstage: Detailing and Testing the Final Concept

Once again, the designers updated their design specification to address the new concerns raised by the nursing staff and to incorporate the requirements and desired characteristics of the new mobile sensory technology solution. Based on this updated design specification, the design team developed a prototype called Feel-It. The students 'negotiated' with the technology in a process which involved programming lights and sound and selecting materials. Once again, this prototype was to serve as a key element in staging a final intervention at the nursing home, and once again members of the nursing home staff were appointed prototype ambassadors.

Over a two-week period, the final prototype was tested in terms of its ability to provide a sense of ease for the residents during stressful situations. In the subsequent feedback session, the nursing staff highlighted that the prototype integrated seamlessly into their existing care practices and was easy to carry around from one room to another. In their opinion, it provided the right type of stimulation for the residents that generated a sense of ease in stressful situations: 'Peter, Hans, and Karen are calm when they have it [the prototype] in their hands and touch it.'

However, the nursing staff also suggested a number of areas for improvement. In some cases, the object created a bit of confusion because it was not familiar to the residents. Physically, the prototype was too big to carry around in a pocket, the sound and vibration length needed to be extended, and the surface needed to be developed in another material to make it lighter and easier to clean, as well as cheap to manufacture. With these modifications, both the nursing staff and the residents would be happy to use this new mobile sensory stimulation technology.

Concluding Reflections on Using Technology in Dementia Care

The focus of this particular project was sensory stimulation technologies, as this was the initial frame and matter of concern expressed by nursing home managers. However, we would like to emphasize that technologies might not necessarily be the most appropriate way to address the challenges experienced by elderly residents with dementia. Such technologies should never stand alone, but be implemented as part of a holistic social–pedagogical approach to nursing care. This is why it was crucially important for the students to engage the nursing staff and help them develop a sense of ownership throughout the design process.

DISCUSSION

First, our case illustrates the large number of material objects produced and staged throughout a design process. At the beginning of the project, the focus was on investigating and interpreting the problem that was going to be solved. This was accomplished by staging negotiations in terms of producing objects such as design games that would represent different framings that could challenge and/or spark negotiations of matters of concern. In this process, the affinity diagram proved particularly interesting, as it continuously evolved and grew throughout the initial phases of the design process and came to represent the concerns of all actors who were engaged in the project. Thus, through negotiations, nursing home managers, nursing staff, and elderly residents with dementia, as well as literature on sensory technologies, etc. were represented on hundreds of post-it notes comprising the affinity diagram. Importantly, the diagram itself grew and was enacted multiple times in negotiations with the design team. It was reworked and used to analyse the empirical data which led to reframing the overall matter of concern from trying to motivate residents and staff to use the current technologies to instead trying to develop a new mobile sensory technology to help ease tensions between residents and staff in one-on-one situations.

The affinity diagram in our case thus functioned as a special type of 'fluid' intermediary object that continued to evolve while remaining backstage. It proved central in staging efforts, as reframing sparked new questions and thus

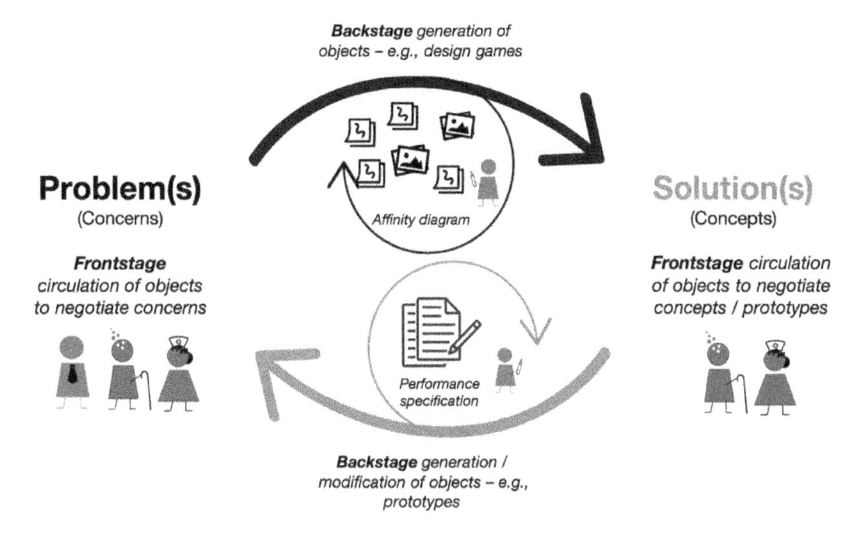

Figure 5.1 *The iterative nature of identifying concerns and framing a solution space*

inspired and gave rise to new objects such as design games to be circulated frontstage in interactions with residents and nursing staff.

As the project evolved and the students shifted their attention toward conceptualizing solutions, another fluid backstage intermediary object was produced and gradually developed. The design specification became the new anchor which was updated based on frontstage negotiations, and which framed the solution space and gave rise to new objects. Potential solutions in the form of prototypes of varying fidelity were circulated frontstage in an iterative process of testing and negotiation that continuously narrowed the solution space. Thus, our case demonstrates how the work of updating the design specification based on expressed concerns leads to framing the solution space (see Figure 5.1), which clearly exemplifies the co-evolution of problem-solution pairs (Dorst and Cross 2001).

CONCLUSION

Adopting the conceptual framework of frontstage and backstage activities, we have investigated how problems (matters of concern) and solutions are negotiated through an iterative process of producing and inscribing objects backstage and circulating these objects frontstage. Our findings contribute to the literature by showing how intermediary objects not only represent processes and solutions, but also represent actors' matters of concern. Furthermore, we have identified a new type of 'fluid' intermediary object that continuously evolves but remains backstage. Objects such as affinity diagrams and design specifications have the potential to represent actors' concerns while simultaneously framing the solution space. In light of these characteristics, such objects merit special attention in the engineering design process, as they play a central role in staging negotiations with multiple actors by inspiring and framing frontstage activities while bridging the gap to the conceptualization of solutions. It should be noted, that in this case example the backstage negotiations are relatively frictionless as the student team is involved in every aspect of the design process. Thus, investigating backstage work in larger organizations involving thousands of actors would require further attention towards the multiple and complex negotiations. In such cases it might be relevant to investigate the staging efforts through a combined theoretical approach of staging negotiation spaces and organizational studies focusing on the staging and formation of action-nets (see Chapter 2 and Chapter 6, this volume).

REFERENCES

Alzheimer's Association (2020), *Alzheimer's Association*, accessed 3 August 2020 at https://www.alz.org/alzheimers-dementia/treatments/treatments-for-behavior.

Boujut, J.-F. and E. Blanco (2003), 'Intermediary objects as a means to foster co-operation', *Computer Supported Cooperative Work*, **12** (2), 205–19.

Brandt, E., J. Messeter and T. Binder (2008), 'Formatting design dialogues—games and participation', *CoDesign*, **4** (1), 51–64.

Bratteteig, T. and J. Gregory (2001), 'Understanding design', in S. Bjørnestad, R.E. Moe, A.I. Mørch and A.L. Opdahl (eds), *Proceedings of the 24th Information Systems Research Seminar in Scandinavia (IRIS 24)*, **3**, Bergen: University of Bergen, accessed 15 May 2020 at http://heim.ifi.uio.no/~tone/Publications/Bratt-greg-01.htm.

Dorst, K. and N. Cross (2001), 'Creativity in the design process: Co-evolution of problem-solution', *Design Studies*, **22** (5), 425–37.

Hendriks, N., L. Huybrechts, A. Wilkinson and K. Slegers (2014), 'Challenges in doing participatory design with people with dementia', in *Proceedings of the 13th Participatory Design Conference on Short Papers, Industry Cases, Workshop Descriptions, Doctoral Consortium Papers, and Keynote Abstracts—PDC '14—Volume 2*, Windhoek, Namibia: ACM, pp. 33–6.

Hendriks, N., K. Slegers and P. Duysburgh (2015), 'Codesign with people living with cognitive or sensory impairments: A case for method stories and uniqueness', *CoDesign*, **11** (1), 70–82.

Latour, B. (1987), *How to Follow Scientists and Engineers through Society*, Cambridge, MA: Harvard University Press.

Latour, B. (1992), 'Where are the missing masses?', in W.E. Bijker and J. Law (eds), *Shaping Technology/Building Society: Studies in Sociotechnical Change*, Cambridge MA: MIT Press, 225–59.

Lindsay, S., K. Brittain, D. Jackson, C. Ladha, K. Ladha and P. Olivier (2012), 'Empathy, participatory design and people with dementia', in *Proceedings of the*

2012 ACM Annual Conference on Human Factors in Computing Systems—CHI '12, Austin, TX: ACM, p. 521.
Olsen, A., A. Didriksen, C. Olsgaard, E. Nilsen and T. Larsen (2016), *Design og Anvendelses af Prototyper. Student Report. Sustainable Design BD3.*, Copenhagen: Aalborg University.
Pedersen, S. (2020), 'Staging negotiation spaces: A co-design framework', *Design Studies*, **68**, 58–81.
Robinson, L., K. Brittain, S. Lindsay, D. Jackson and P. Olivier (2009), 'Keeping in Touch Everyday (KITE) project: Developing assistive technologies with people with dementia and their carers to promote independence', *International Psychogeriatrics*, **21** (3), 494–502.
Rodgers, P.A. (2018), 'Co-designing with people living with dementia', *CoDesign*, **14** (3), 188–202.
Vinck, D. (2012), 'Accessing material culture by following intermediary objects', in L. Naidoo (ed.), *An Ethnography of Global Landscapes and Corridors*, Rijeka, Croatia: InTech, pp. 89–108.

PART III

Staging changes in networks and organizations through design of spaces and events

6. Staging the configuration of organizations for social innovation impacts

Jens Dorland

INTRODUCTION

The developed world is facing numerous challenges to the modern welfare state. Social innovation organizations seek to overcome these challenges by addressing societal issues such as social justice, inequality, changing demographics, environmental preservation, and climate change (Dorland 2018). The focus in social innovation, broadly understood as addressing a specific type of challenge, typically centres on finding solutions to societal problems; few have examined how organizational dynamics affect the ability to develop and implement such solutions. Here, I explore how a staging perspective can be used to structure organizations for maximum societal impact.

Many social innovation organizations and initiatives began as informal, disparate and distributed activities that gradually became more coordinated through interactions and developing social relations (Dorland 2018). What begins as a series of disconnected events and informal meetings develops into more formal events such as workshops and conferences that gradually become connected through staging and sensemaking processes to create action-nets that constitute new organizations. The initial hurdle of creating a space or scene where this process can take place and continue to develop is especially challenging for social innovation initiatives and their networks because they lack the resources and natural boundaries of businesses and public institutions.

In this chapter, I present case studies of two social innovation networks: the Living Knowledge Network (LKN) and the INFORSE renewable energy movement. Focusing on one local initiative in each network, I examine how staging structured organizational dynamics for maximum impact. Findings from these case studies reveal fruitful strategies that can be implemented by both policymakers and practitioners to support the emergence and development of organizations that contribute to solving societal problems.

A PROCESS AND ACTIVITY VIEW OF ORGANIZATIONS

Inspired by scholars such as Weick, Goffman, Czarniawska, and Latour, a process view of the organization comprises an interesting marriage of actor–network theory (ANT) and organization studies, depicted in Figure 6.1. Because material semiotics and related approaches such as ANT are theories of social order and not specifically of organizations (Callon and Latour 2015; Czarniawska and Hernes 2005), I am equally inspired by organization theory (Czarniawska 2006; Czarniawska and Hernes 2005; Hernes 2008). ANT has been widely applied in studies of organizations, but as noted by Czarniawska and Hernes (2005), an aporia of translations exists. In their anthology, they tried to narrowly adapt ANT to the specific area of organizational studies and an approach 'that makes it possible to show how power emerges through organizing' (Czarniawska and Hernes 2005: 10). The agenda in this chapter is to empower civil society organizations such as energy cooperatives in INFORSE by showing how they can 'construct' power.

From this perspective, organizations are viewed as a process to be identified by examining how entities organize, and ontologically are brought into being as organizing activities are performed (Robichaud and Cooren 2013). In addition to temporality that the process perspective considers, I will emphasize material semiotics, which is also a focus in ANT: underexposed material aspects such as the 'props' used on Goffman's (1959) dramaturgical stage, and the inanimate objects travelling through time from one event to the next which have agency or serve as agents for transferring agency. This view can be traced back to Weick (1995), who further drew inspiration from Goffman, Allport, and other scholars who have studied the sociology of interaction (Czarniawska 2006). Weick (1995), however, took it further and saw strings of events as a process of collective sensemaking that constituted an organization, although from a linguistic perspective.

Sensemaking is the process whereby people collectively give meaning to their experiences by connecting past moments with present experience (Weick 1995). Over time, this process yields what Czarniawska (2006) termed action-nets: structures of associations between actors that are stabilized through objects produced by events. Figure 6.1 therefore depicts both the process and the event, which have an effect on the existing action-net through objects. So, it is the events that form the structure and have agency and thus the actions themselves that are the actors, hence the term action-nets. Action-nets foreground the actions and have a more explicit temporal perspective—that is, the actions and events that occur at specific times and the objects they produce are what forms continually evolving organizational structures, and the

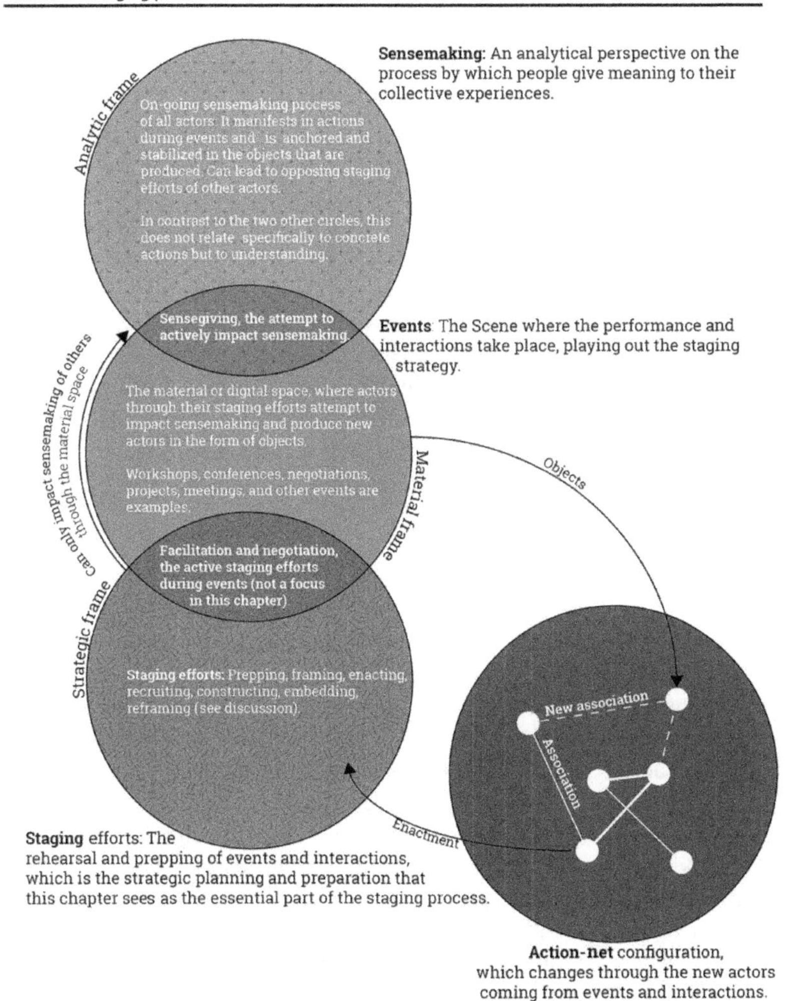

Note: Visualization of the theoretical staging framework, with the main conceptual terms highlighted in bold. The focus in this chapter is on the strategic and preparatory staging efforts that bear fruits during events, resulting in objects that can lead to specific configurations of action-nets. The details of what happens during events and the facilitation is blackboxed, as well as the sensemaking process of other actors. The scene, or the event, is essential, as all relations and knowledge need to be materially anchored in the world not to be ephemeral and to be able to travel (Czarniawska-Joerges and Sevón 2005).

Figure 6.1 Staging framework based on events, objects and action-nets

objects allow events to have an effect across time and space. This is identical to ANT as laid out by Latour (2010), where he focuses on the moves and not the 'thing', i.e. what seem like stable organizations. Action-net is thus a better term, also to describe Latour's perspective, than actor-networks.

Importantly, the sensemaking perspective is very analytical and not action-oriented, and sensemaking is not an overt activity. However, the focus on individuals is empowering, as it informs individuals of potential actions and their consequences. Or, as put by Czarniawska and Hernes (2005: 12) 'ANT can empower micro actors, explaining to them the mechanisms of power creation'. Drawing on Weick gives the approach more explanatory weight, whereas ANT can tend to become very descriptive, making it harder to use to inform staging approaches. The originality of this chapter is the focus on developing staging as a tool that attempts to have intentional impacts on unfolding events, in contrast to more passive interpretations of historical events in many studies. The process view of organizations is especially suitable for this agenda. In contrast to some of the other chapters in this book that focus on navigation during events, this chapter focuses more on the activities taking place between events.

STAGING THE DEVELOPMENT AND EMERGENCE OF ORGANIZATIONS

Staging in this chapter refers to the strategic intentionality in creating or facilitating organizing efforts. Goffman (1959) understood staging as a social actor that is constantly concerned with ensuring the success of communication, interaction and cooperation by means of mobilizing and drawing together many different actors such as texts, scripts, narratives, supporters and so forth. An important aspect is the separation between the backstage, where strategic planning takes place, and the frontstage where the staging strategy plays out and sensemaking occurs. In the cases presented in this chapter, staging relates to the activities occurring between events, such as the development of a manuscript for a workshop, and the construction of textual artefacts and media objects to be enacted during the workshop. In this way, staging is an act of trying to configure an action-net in a specific way. As the cases demonstrate, sequences of interactions produce objects, roles and texts that can be enacted in later interactions while also shaping them, thereby forming an action-net. Such action-nets can be perceived of as organizations, constantly evolving and developing, if they have enough stability for other actors to observe and interact with them. These organizations are the 'stable' structures that actors recognize, or the 'scene' when drawing on the theatrical metaphor, which can be the object of scrutiny and political action.

METHODOLOGY

This chapter is based on two of the 20 case studies and the meta-analysis comprising the TRANSIT project (www.transitsocialinnovation.eu). The LKN and INFORSE cases involved several local initiatives; here, I focus specifically on the Videnskabsbutikken and Samsoe initiatives due to their long histories. I personally did the case study on LKN (Dorland and Jørgensen 2016), whereas colleagues produced the case study on INFORSE (Elle et al. 2015). The case studies were based on semi-structured interviews, participant observations and document analysis. See Dorland (2018) for details on data sources and methodology. A public database of Critical Turning Points (CTPs) (Pel et al. 2017) was created in the meta-analysis that organized the data from the two cases into chronological events; quotes for the Samsoe case were taken from this database (see http://www.transitsocialinnovation.eu/sii/inforse-2).

CASE 1: VIDENSKABSBUTIKKEN AND LKN

The Science Shop (ScS) concept was developed in the 1970s at Dutch universities in response to growing demand from citizens and civil society organizations (CSOs) for access to research. The ScS model challenged the traditional orientation of science with regard to how knowledge is developed and strengthened the influence of communities on research. In this section, I examine a local ScS initiative called Videnskabsbutikken (VB) at the Technical University of Denmark (DTU).

The VB case exemplifies how staging can be used to establish a new scene or space inside a large institution, where the actors could work to stabilize the action-net that would become VB. The empirics are ordered as subsequent episodes shown in Table 6.1 to help analytically string together an action-net revealing how VB developed over time. Interestingly, the staging efforts in this case mostly focus on staging objects of different kinds, and very little staging in relation to people. In this section, I describe two episodes from Table 6.1 in detail.

Episode 1: Pilot Project

Episode 1 began with a request in 1983 from a labour union to explore how the equipment, competences and employees from a closing shipyard might be used. This unplanned event led to collective sensemaking among members of several groups at DTU, who framed it as demonstrating a need for a more formalized open door for CSOs such as labour unions, as well as strong demand for research, ideas that were well-received by the university at the time.

Table 6.1 *Major chronological events for the VB case, based on a more detailed table of 25 events*

Episode	Events	Outcome	Staging efforts
Episode 1: pilot project (1983–1985)	The closure of a shipyard was used to create a narrative supporting the creation of VB. Students were sent to study the ScS concept in the Netherlands; the concept was used to further legitimize the initiative. A long process of meetings and negotiations at the university followed, resulting in the submission of a project proposal to the academic senate.	VB was founded as a three-year pilot project	Creating a narrative enacting local events Enacting known concepts Showing student engagement Enacting societal discourse, social responsibility of universities Negotiating with local actors (professors from a range of departments), to enrol them as allies and ambassadors
Episode 2: expansion (1986–1995)	The coordinator of VB engaged in staging to make the organization permanent. He actively encouraged CSOs to request help, and compiled statistics about the number of requests submitted; he also created artefacts to promote successful projects, thereby constructing objects to support VB.	VB became a permanent organization within the university	Constructing objects that can subsequently be enacted Prepping CSOs to make requests for student projects

Episode	Events	Outcome	Staging efforts
Episode 3: founding LKN (1997–2001)	From the mid-1990s, VB was in decline due to cuts in funding and the disintegration of its research group, which led to a search for new allies and resources. LKN was born out of interactions with other ScSs in the European Union (EU) that led to an EU-funded project that provided enough resources and a scene to develop LKN.	LKN was formally founded as an ScS network	Constructing objects that documented impacts on research and education Creating alternative roles that could be enacted locally Affecting sensemaking at the university by demonstrating wider international support for the concept Defining the ScS concept more formally to prevent changes
Episode 4: modern era (2000–2012)	VB managed to survive thanks to various staging efforts during a long period of decline. LKN provided some resources through international projects, and projects with municipalities helped reframe the narrative while showcasing student projects. VB also formed an alliance with the external communications office and formally became a sub-unit responsible for civil society requests.	VB survived, but formally closed when the coordinator changed jobs	Staging a new sensemaking of VB by creating new narratives Enacting student projects to create new narratives Recruiting new allies Connecting to other action-nets to increase stability

Source: See Dorland 2018: 212.

Members of the groups who produced this framing were politically active students and faculty members who wanted to help solve societal problems, henceforth termed activists. Around the same time, the activists encountered and embraced the ScS concept, also inspired by the alternative technology movement (Smith 2005). The ScS concept thus became one of the objects that were enacted together with the story of the shipyard in a staging process to establish VB. Here, staging with the intention to facilitate collective sensemaking was essential, as it established alignment between university management and the activists. This alignment was also possible at the time because management constituted a senate of academics open to discussion and negotiation. These objects, together with a show of broad interdisciplinary support from several institutes enabled VB to obtain funding for two positions.

There is a distinction between the organization the activists tried to create (VB) and the scene of this initial staging process. The university was the stage where the negotiations took place, defined based on its role in producing knowledge to solve societal problems. In the staging process, actors outside the university were enacted through objects created by the activists like the narrative of the shipyard closing, but were not active participants in the performance, as the initial staging process was an internal negotiation for resources at the university. The staging process constructed an action-net that could function as a scene for the organizing efforts of VB that connected events spanning the university and broader society, thereby enacting these outside actors and wielding their agency in the staging efforts to affect the internal sensemaking process at the university. Through the action-net that the staging process constructed and stabilized VB became a recognized organization within the university with funding, an office, staff and a defined purpose. Texts featured prominently as actors in this process, representing earlier events, stabilizing the action-net and documenting outcomes in the form of contracts. The prevalence of text is also due to the nature of the academic context and is not as prevalent among other cases on social innovation.

Episode 3: Founding LKN

The transition to the twenty-first century was a turning-point, with the university governance system changing from a university senate of academics to a government-appointed rector, thereby recasting the actors on the stage. This was accompanied by a shift in university policy that emphasized the commercialization of science. Relative to a senate, a rector-led management structure involved less negotiation, resulting in a lower likelihood of collective sensemaking. The new rector had radically different priorities, which essentially rendered the action-net of VB ineffective, as many of the objects lost power, such as the narrative around helping CSOs, the ScS concept, documentation of student projects and so forth. New narratives and objects had to be created and enacted that were more in line with the new sensemaking of the university management.

The establishment of LKN and EU projects in subsequent years led to funding for new projects and staff replacing some of the lost resources. The first EU project acted as a scene or space where the local ScSs could establish closer relations with each other, learn how best to operate, develop infrastructure and engage in collective sensemaking. In this project text objects documenting and framing the value of ScSs were constructed, which would later form part of the action-net of LKN. A key milestone for LKN was the final conference in the EU project that was a necessary physical space that together with various project documents enabled staging efforts to formally establish the organization of LKN. After the establishment the network provided new roles as well, as ScSs became local representatives of an international network, which really meant something locally (Pel et al. 2017). The organizing efforts of LKN enabled interaction with institutions in the EU, and over time enabled a sensemaking process with the commission that resulted in ScSs and CSOs becoming part of the various research programmes and frameworks, which ScSs could then enact at their universities.

CASE 2: SAMSOE, DENMARK'S FIRST SUSTAINABLE ENERGY ISLAND

Established in 1992, INFORSE is an non-governmental organization (NGO) network promoting sustainable energy development based on renewable energy, energy savings and social development. It comprises 140 independent NGOs working in nearly 60 countries. In this section, I focus on a local initiative to establish sustainable energy on an island in Denmark, a project that became the Samsoe Energy Academy, henceforth called Samsoe. Table 6.2 orders the empirics of the case in chronological order structured around events; in this chapter I will focus on events 1, 2, and 5.

Episode 1: Initial Proposal

The origins of the energy community on Samsoe can be traced to a local socio-economic crisis, the closure of the local slaughterhouse, similar to the closing of a shipyard in the VB case. In response, several key actors on the island began to consider the future and to discuss whether they should shift their focus toward agricultural products, tourism or something else. This event coincided with the establishment of a green fund by the Danish government leading to the idea of becoming the first sustainable island in Denmark.

Table 6.2 Major chronological events for the Samsoe case

Episode	Events	Outcome	Staging efforts
Episode 1: initial proposal (1997–1999)	A slaughterhouse that was a major employer on the island closed. The ministry established a fund for sustainability projects.	Samsoe submitted a proposal for a project to become the first sustainable Danish island	Framing the project as a response to the closure Enacting the fund to facilitate sensemaking about the focus for the project Embedding concerns
Episode 2: learning public involvement processes (1998–2001)	A range of informal meetings and public events enabled founders to learn how to involve the community and stage interactions.	The sustainable island project was successful	Prepping public meetings Recruiting allies Arranging events
Episode 3: creating a meeting place with content (1999–2007)	The founders constructed objects from the knowledge generated in the project, such as documentation and learning materials, to brand the island and attract eco-tourists.	The organization accumulated objects over time	Constructing objects
Episode 4: lack of funding and new opportunities (2001)	Changes in the context led to funding cuts for renewable energy projects, leading to a reframing of the project into energy savings.	The project changed focus to energy savings through district heating, leading to a material transformation of the infrastructure	Reframing the project Recruiting allies, prepping craftsmen to act as 'salesmen' for the project Embedding concerns
Episode 5: energy academy on Samsoe Island (2007–)	The project and the local energy office establish the energy academy as a formal organization.	The energy academy was established, stabilizing the action-net	Framing local challenges, the future of the island Prepping and arranging events Constructing allies

There were strategic considerations from the start. The actor who proposed the idea was a 'newcomer' who had only lived there for ten years, so she recruited a local resident with a family history on the island who was 'able to talk to "the tribe"'. This led to a workshop on 'good energy' with 150 participants designed to attract support from the residents by framing their own dreams in relation to the larger context of the island. They also set up a conference on social action with the laid-off slaughterhouse workers, the labour union and other actors as part of the co-creation process aimed at designing the future of the island. This conference was divided into two parts: a closed workshop with the most 'powerful' residents, followed by an open workshop. The conference was structured this way to enlist support from the 'elite', several of whom refused to mingle with other residents. Proponents of the project succeeded and received support from residents and funding from the government.

Episode 2: Learning Public Involvement Processes

Following this success, the project managers developed a master plan adopting a technical approach to address all engineering and structural questions. However, these issues were not what concerned the residents, who focused on ownership and on who the project would benefit and how. The initiators thus learned that they needed to facilitate buy-in throughout the process:

> It is not the meeting itself that is the most important; it is the discussions before and after the meeting. In order to have a successful meeting you have to prepare well, be sure that your important allies will be present at the meeting ... We learned how to 'prime' a meeting. We learned that we had to walk from house to house, discussing the issues of the meeting, making sure that supporters would turn up at the meeting. (Quote from CTP database, see Pel et al. 2017)

This illustrates that a lot of the activities involved in staging take place between events. Project leaders worked to ensure that allies would turn up and that sensemaking was in sync, which enabled them to focus the agency of a network of actors to secure agreement on issues of relevance and related actions. One of the implications is that a lot of negotiation and decision making does not take place at events (meetings, workshops, conferences, etc.).

Episode 5: Energy Academy on Samsoe Island

Episode 5 pertains to the establishment of an energy academy and a strategy dubbed Samsoe 2.0. The energy academy was established when the action-net of the ongoing project was formalized as an organization, a more stable and permanent action-net for containing the organizing process. The academy was intended to be permanent and to grant visitors access to knowledge generated by the initiative. This conversion coincided with an event in 2007 when they achieved carbon neutrality. After achieving this longstanding goal, the question became: 'What next?' The initiators had a goal of becoming fossil-free as well, and they began to develop a new strategy:

> It was essential to have all the local people in power participating in the process of Samsoe 2.0. We had to design a process involving these people and the rest of the islanders. We started with a shared space with 25 of the most powerful people on the island—this event was filmed. The event was planned and designed in every detail.

> After this we had a banquet to which we had invited all the islanders—but made it possible for some of the people who had participated in the shared space event to leave. Some of the powerful people did not like the idea of mingling with the rest of the islanders. (Quote from CTP database)

The final strategy also involved other actions not related to sustainability, such as establishing better infrastructure for information and communications technology, and fighting for a new ferry. These inclusions were part of attracting support for the plan, recruiting allies. The initiators also framed the question of the workshop as 'What will happen, if we do NOT engage in further change?' Framing and enacting the agenda in this way convinced critical voices that inaction was more expensive.

THE STAGING OF ORGANIZATIONS FOR SOCIAL INNOVATION

My analysis reveals many similarities between the two cases. The origins of both initiatives can be traced to an unexpected event followed by informal interactions among people with an ambition to address related societal problems in their local context, which led to a series of strategically staged events that over time formed an action-net supporting these organizing efforts. Ultimately, both initiatives led to the establishment of formal organizations (However, VB was dissolved in 2012 when the founder switched jobs). The question is: What can we learn about staging the establishment of an organization in a way that empowers actors and enables them to have maximum societal impact?

In the first case, innovation aimed at solving societal problems was enabled through university–civil society interactions by stabilizing the action-net that defined VB as a distinct organizational unit within DTU. In the second case, the act of establishing an energy cooperative on Samsoe stabilized and formalized the action-net of the project, wherein the organizing process and staging efforts that facilitated a fundamental transformation of the island community could continue. The initiators of both cases strategically enacted past events—the closure of the shipyard and the slaughterhouse—to create narratives that supported their agendas. VB was aligned with the broader sensemaking of the university at the time, and academics had more power and influence, making the narrative they constructed effective along with broad enrolment of different institutes and students. Samsoe strategically engaged 'locals' and craftsmen as ambassadors who could facilitate enrolment of the wider community.

As illustrated in the different episodes, the staging efforts in the cases were (in loose chronological order):

- **Framing**: Creating supporting narratives by framing current events in their local contexts as problems their proposed initiatives could solve.
- **Enactment**: Enacting public policy and societal discourses to support the narratives.
- **Recruiting**: Recruiting allies with influence and establishing supporting networks.
- **Prepping**: Preparing for events by ensuring allies attended events and supported the planned scripts.
- **Constructing**: Constructing objects such as documentation based on statistics and scientific methods.
- **Embedding**: Embedding concerns of other actors through participation and co-design to create the conditions for collective sensemaking—making the action-net a meaning frame.
- **Reframing**: Reframing the narratives of the initiatives to align with contextual changes.
- **Events**: Arranging events such as conferences and workshops that produced objects that could be enacted in the future.

These staging efforts largely take place between events but aimed at constructing objects during events and interactions to focus the distributed agency of the participants for later enactment and configuration into action-nets. Prepping an event is a strategic staging effort to control the action, which can be enacted to later configure the structure to support the initiative. In both cases, the initial staging efforts resulted in seed money that was used to stabilize the action-nets they had established. The rest of the chapter illustrates some of these staging efforts in relation to different objectives.

The Staging Process and Objectives

Staging to establish an organization

In both cases, the initiators facilitated a continuous iterative staging process, first to create the action-net of their developing organizations, and then to stabilize it. Narratives are especially important objects to create a story that frames the necessity and relevance of the proposed initiatives.

Likewise, both initiatives were started by framing socio-economic crises through interactive and participatory processes to facilitate collective sensemaking. Either the concerns of other stakeholders need to be negotiated and embedded into the narrative, or their sensemaking of the initiatives needs to be changed so they can see how their concerns are intertwined. These can be described as processes of collective sensemaking on one hand, and of sensegiving (Weick 1995) on the other, where sensegiving is more unidirectional.

VB project leaders facilitated sensegiving to connect management's perspective on the purpose of the organization to the purpose and role of the university. At that time, helping labour unions was viewed as a favourable endeavour. For Samsoe, collective sensemaking took place in workshops, where participants put their dreams and visions on post-it notes on the walls, and project leaders framed these in the context of the island and the initiative to show how their concerns would be addressed. In this way, the vision of Samsoe as a sustainable energy island was developed from the bottom-up and collective sensemaking achieved through embedding.

Staging for survival

Both initiatives faced crisis and opposition related to loss of funding as a result of government and policy changes. The action-net that had created and stabilized VB changed significantly as the university reframed what counts as innovation to focus on commercialization. The narrative of helping CSOs gradually lost traction. Beginning in the mid-1990s VB faced a series of challenges and responded in different ways, by: trying secure resources from elsewhere through international networking, in terms of both funding and various roles and objects that could be enacted; developing new narratives by partnering with a local municipality; and engaging in strategic staging efforts with the intention of sensegiving at the university by constructing various objects, such as news stories for the university paper. VB was able to adapt because the coordinator had undergone a sensemaking process that made him realize that it was essential to tell new stories in the changing context of the university.

For Samsoe, the funding and subsidies for renewable energy were cut, so project leaders reframed the sustainable island project to focus on energy savings and district heating. The conversion to district heating would create work for several groups of craftsmen, thereby making it possible to enlist them as ambassadors for the project. Initially, the project was framed in relation to sustainability, but they quickly learned to emphasize the increased comfort and cost savings of district heating in the framing to embed the local concerns.

Staging for stabilization

The achievement of carbon neutrality was a turning-point for Samsoe. This could have marked the end of the project as this was the initial goal. However, the founders staged the creation of a new vision, Samsoe 2.0, spanning the next 20 years. They negotiated with the community and included some aspects that were unrelated to sustainability to ensure their support. Through a series of meetings and a workshop, they came up with a solution to convert the local energy office into an energy academy that would be open to visitors. The founders had become more proficient at staging, and this time documented the workshop through a video as a way to stabilize the outcome. The energy academy was a more stable action-net that could serve as a base of operations, as it was formalized, permanent and not contingent on project timelines.

PRODUCTIVE STRATEGIES FOR STAGING ORGANIZATIONS FOR SOCIAL INNOVATION

Insights from these cases can be used to inform productive strategies for staging the establishment of communities and organizations to facilitate social innovation. I have identified four distinct strategies that might be implemented sequentially or individually. As illustrated by the cases, iteration amongst these strategies is necessary to ensure survival and maximum impact.

Organizing a Scene for Sensemaking

For collective sensemaking to happen, it is crucial to have a scene on which to meet. This scene can be physical or virtual, as long as it facilitates interaction. This scene can be staged to impact sensemaking in different ways. Although sensemaking is a necessary process for network creation (Dorland 2018), it is important to identify whether the objective is to facilitate sensemaking in a genuine collaborative effort, or to engage in sensegiving to impact the sensemaking of others in a certain way. Once sensemaking has taken place and an organization begins to emerge, i.e. the process of organizing is ongoing, it is important to establish a backstage for strategy development and more overt staging activities such as the construction of objects that can later be enacted to give strength and stability in an action-net.

This strategy was employed in both cases. VB helped stage the creation of a new action-net for the wider international network that developed into LKN, which in turn played an essential role in constructing objects that could be enacted to protect and stabilize VB in the local context. For Samsoe, project leaders established an action-net through the project where the island could develop as an energy community and enabled the continuous enrolment of new actors.

Constructing Allies

Once an organization starts emerging it is important to construct objects for later enactment. In other words, the action-net must be expanded and stabilized. The construction of objects is a way to focus the dispersed agency of the actors behind it to wield their power.

In the first case, the coordinator spent time documenting the demand and success of VB in episode 2. The coordinator also produced reports documenting VB's positive impact on research and education at universities in episode 3. LKN also successfully interacted with the EU commission to convince them to incorporate community research and ScS principles into future research frameworks that could later be enacted across universities in the EU. Samsoe likewise constructed allies. Project leaders not only documented their progress toward carbon neutrality, energy savings and the increasing number of eco-tourists, but also used the workshops and meetings to make videos and other documents such as the Samsoe 2.0 vision, which legitimately represented the entire island as a community.

Enacting Objects to Gain Agency

The objects and roles described previously can be enacted to establish agency. In the VB case, several events were framed and enacted in the project proposal in episode 1. Later, VB enacted student projects in various ways to demonstrate impact and increase visibility. Samsoe used documentation in various ways to show cost savings, promote success stories, or highlight economic development potential of tourism, which among others led to the formalization of the project into a more recognizable organization by converting the local energy office into the energy academy, thereby increasing the stability.

Facilitating Collective Sensemaking

Collective sensemaking in a supportive context differs radically from that in the face of opposition. In this alternative context, sensegiving is necessary to facilitate collective sensemaking. Initially, VB operated in a supportive context, thus sensegiving was not necessary. Likewise, the government initially supported Samsoe's focus on sustainability and renewable energy through funding; however, the local context was more contentious, and sensegiving work was necessary.

In later years, the sensemaking of the DTU management changed, resulting in a disconnect with VB. In response, VB tried to impact the sensemaking at the university by framing itself in relation to DTU's new priorities, such as partnerships with public authorities. This sensegiving strategy was intended to re-establish legitimacy without VB having to change its internal sensemaking, i.e. VB tried to hide the contradiction by engaging in sensegiving, thereby enabling the organization to survive and pursue its current activities. This organizational contradiction is missing from the process view of organizations, especially Weick's conceptualization of sensemaking. For Samsoe the national context became less supportive while the local context became more supportive. This local support enabled them to stage the sustainable island project as an energy-savings agenda. As these cases demonstrate, the degree of support and opposition is a fluid dimension and achieving collective sensemaking or consensus is not always a given or a necessity.

CONCLUSION

In this chapter, I have developed a framework of sensemaking and staging in an attempt to build a heuristic to explain how different staging efforts facilitate and influence sensemaking and alignment with other actors to create organizations understood as action-nets. Action-nets focus the dispersed agency of actors through events, which among other things can construct objects with power. In the cases presented here, the founders of VB and Samsoe used such objects to create and stabilize their organizations.

Typically, the staging concept from design studies (Clausen and Gunn 2015; Clausen and Yoshinaka 2007) is applied to development processes of products and technology; here, I have extended the concept to organizational studies by applying it to cases of two social innovation initiatives. Combining the process view on organizations with the event focus of Weick and a material semiotics perspective enables an understanding of how events develop into action-nets. Staging thus unfolds impactful events. As organizations are understood as processes, and staging likewise is an organizing process, the implication is that organizational processes structure organizational processes in a never-ending iteration. Staging, however, is the overt, intentional, and strategic attempt at affecting the ongoing organizing process. Staging provides a more action-oriented design perspective on how to actively establish and alter organizational structures, the temporarily stable action-nets, by creating objects through events such as workshops.

Staging is in this chapter thus viewed as an overt organizing effort, whereas sensemaking is related to the process taking place within the conditions created through staging. This chapter contributes to the literature by elaborating the distinction between overt strategy that staging entails and the sensemaking process.

I applied this framework to two cases. The VB case illustrates how social innovation organizations can use staging to construct text objects that can create agency and support the action-net of an organization, enabling the continuous organizing and staging efforts. The sustainable island project on Samsoe focused on a sustainable transition of the island, as well as economic development, and illustrated the importance of adopting a participatory approach to staging and how to adapt the framing of a project. Both projects required scenes where community members could initially meet to develop ideas and enable collective sensemaking, and later engage in strategic staging efforts. Based on my analysis, I was able to identify four fruitful staging strategies that can be implemented by both policymakers and practitioners to support the emergence and development of organizations that contribute to solving societal problems.

- **Creating a scene for sensemaking**: Collective sensemaking is necessary to establish an organization that can wield the agency of a network. However, sensemaking needs to have a scene on which to unfold. Finding or creating a scene for collective sensemaking and enabling further development of an action-net is an iterative process.
- **Constructing allies and power**: Action-nets are largely composed of objects from past events. During events, it is crucial to construct objects that can be enacted later, such as scientific documentation, media coverage, signed memoranda of understanding, reports, vision documents and so forth, which focus the dispersed power of the actors behind the objects. It is necessary to engage in reflection and strategic planning to determine which objects will be needed and what forms they should take.
- **Enacting objects to gain agency**: The power embedded in objects needs to be enacted in interactions to provide agency. It is important to identify which arguments and objects are accepted in each interaction, as well as the priorities, goals, and motivations that structure and limit how objects can be enacted.
- **Facilitating sensegiving or collective sensemaking**: Scenes must be staged to facilitate collective sensemaking through interactions and negotiations, or sensegiving to influence the sensemaking of others by enacting objects, thus aligning the organization with the sensemaking of relevant stakeholders.

As illustrated in the case studies, it is important to recognize that many staging activities take place *between* events, rehearsals for the play to come. Events are thus the culmination of a staging process, hopefully leading to tangible outcomes that can be used in developing an action-net stable enough to be perceived as an organization. Such action-nets are the frame for the organizing that in the process view on organizations *is* the organization.

REFERENCES

Callon, M. and B. Latour (2015), 'Unscrewing the big Leviathan: how actors macro-structure reality and how sociologists help them to do so', in K. Knorr-Cetina and A.V. Cicourel (eds), *Advances in Social Theory and Methodology*, New York & London: Routledge, pp. 277–303.

Clausen, C. and W. Gunn (2015), 'From the social shaping of technology to the staging of temporary spaces of innovation – A case of participatory innovation', *Science and Technology Studies*, **28** (1), 73–94.

Clausen, C. and Y. Yoshinaka (2007), 'Staging socio-technical spaces: translating across boundaries in design', *Journal of Design Research*, **6** (1–2), 61–78.

Czarniawska, B. (2006), 'A golden braid: Allport, Goffman, Weick', *Organization Studies*, **27** (11), 1661–74.

Czarniawska, B. and T. Hernes (2005), *Actor–Network Theory and Organizing*, Malmo: Liber.

Czarniawska-Joerges, B. and G. Sevón (2005), *Global Ideas: How Ideas, Objects and Practices Travel in a Global Economy*, Copenhagen: Copenhagen Business School Press.

Dorland, J. (2018), *Transformative Social Innovation Theory: Spaces & Places for Social Change*, Aalborg: Aalborg University Press.

Dorland, J. and M.S. Jørgensen (2016), *WP4—CASE STUDY Report: Living Knowledge*, TRANSIT: EU SSH.2013.3.2-1 Grant agreement no 613169.

Elle, M., V. van Gameren, B. Pel, H.K. Aagaard and M.S. Jøgensen (2015), *WP4—CASE STUDY Report: INFORSE.*, TRANSIT: EU SSH.2013.3.2-1 Grant agreement no: 613169.

Goffman, E. (1959), *The Presentation of Self in Everyday Life*, New York: Anchor Books.

Hernes, T. (2008), *Understanding Organization as Process: Theory for a Tangled World*, London & New York: Routledge.

Latour, B. (2010), 'Coming out as a philosopher', *Social Studies of Science*, **40** (4), 599–608.

Pel, B., A. Dumitru, R. Kemp, A. Haxeltine, M.S. Jørgensen, F. Avelino, I. Kunze, J. Dorland, J.M. Wittmayer and T. Bauler (2017), *TRANSIT WP5 D5.4—Synthesis Report: Meta-Analysis of Critical Turning Points in TSI*, TRANSIT: EU SSH.2013.3.3.2-1 Grant agreement no: 613169.

Robichaud, D. and F. Cooren (2013), *Organization and Organizing : Materiality, Agency, and Discourse*, London & New York: Routledge.

Smith, A. (2005), 'Environmental movements and innovation: From alternative technology to hollow technology', *Human Ecology Review*, **12** (2), 106–19.

Weick, K.E. (1995), *Sensemaking in Organizations*, Thousand Oaks, CA: Sage Publications.

7. Staging participatory innovation as transition design

Christian Clausen and Wendy Gunn

STAGING SUSTAINABLE TRANSITION

Our chapter examines a case of participatory innovation as an instance of staging a transition process where more sustainable indoor climate practices have been promoted.[1] Transition and the related innovative processes increasingly have to take place across system boundaries including sequential intersections of design and use. The research discussed here focuses on a 'temporary space' for participatory production of knowledge across design and use, which is both transitory and improvisational in character. We highlight the staging and potential role of such a temporary space in transitioning indoor climate knowledge and practices across engineering systems of design and use.

The chapter attends to the role of intermediary objects in the movement of knowledge via participatory innovation, which emphasizes particular knowledge objects and practices. We draw upon action-oriented approaches to social shaping of technology with particular attention to their discursive and political dimensions as described in Chapter 2 (this volume) exemplified with the analytical notion of sociotechnical spaces for innovation (Clausen and Koch 2002; Clausen and Yoshinaka 2007). The notion of design labs (Binder and Brandt 2008) is incorporated as a way of addressing the staging of a temporary space, which involves the mobilization of research findings from juxtaposing field inquiries and design practices.

Staging of participatory innovation practices is illustrated and analysed through a case study concerned with social shaping of indoor climate conceptions and related technical solutions. Shove (2003), Jaffari and Matthews (2009) and others have presented indoor climate practices as an important but difficult case for sustainable transitions due to path dependent developments sustained by dominant sociotechnical regimes. While we do not present a case of regime transition, we have been looking for new lines of inquiry concerning conceptions of user practice and whether these could lead towards a potential reframing of the way the social shaping of technical indoor climate solutions is being constituted.

Research involved collaboration between the TempoS[2] project on Performing Temporary Spaces for User Driven Innovation at Aalborg University and SPIRE's[3] research on participatory innovation at University of Southern Denmark (Buur and Matthews 2008; Buur 2012; Gunn and Clausen 2013). The aim of this particular collaboration was to trace and investigate the travel, translation and uptake of user knowledge about everyday indoor climate practices, via a series of participatory innovation workshops, into industrial organizations, engineering indoor climate research institutions, and engineering practices.

Our aim is to investigate how the movement of knowledge from local indoor climate practices to the worlds of engineering in the building industry and indoor climate research institutions may be facilitated. Our question is concerned with the staging and politics of the design of temporary spaces spanning design and use and whether these may lead to the reframing of user conceptions and their uptake in industrial and research organizations as a way of enabling transition. To be more precise, we are concerned with how the production of specialized engineering knowledge concerning indoor climate can draw on a more participatory production of knowledge (Tonkinwise 2017). A key issue underpinning the chapter is a concern with the conditions within which engineers make sense of knowledge generated by users.

DESIGNING FOR TRANSITION

Transition is generally concerned with the long-term perspective of transforming sociotechnical regimes such as interlinked patterns of artefacts, institutions, practices, rules and norms (Berkhout et al. 2004). Regime transformation has traditionally emphasized the role of strategic or technological 'niches', a protected space for changing norms or radical innovations forming the key locus from where the regime may be challenged. As pointed out by Geels (2002), whether such innovations lead to regime transitions depends on the alignment and coordination of different actor groups and the possibility of loosening up these linkages. Recently, several authors have pointed at design as a source for transition but that the existing design and innovation approaches do not serve the purposes of systemic transformation (Gaziulusoy and Brezet 2015). Further, transition design has been coined as 'an attempt to resituate the practice of designing within a commitment to facilitate social change towards more sustainable futures' (Tonkinwise 2015: 86). Transition design is here linked to system level or paradigmatic change (Irwin et al. 2015) including changes at all levels of society, everyday practices and long-term structural transition.

The transition perspective adopted in this chapter draws on a social shaping perspective on the staging of design processes (see Chapter 2, this volume). The concept of sociotechnical spaces (Clausen and Koch 2002; Clausen and Yoshinaka 2005, 2007) encompass sociomaterial, political and discursive practices and emerging configurations of sociotechnical ensembles and networks. By bringing attention to political processes in the creation of boundaries delimitating but also enabling certain innovation processes, the notion of the sociotechnical spaces provides a sensitizing device for studies of the staging of innovation, transition and reflexive action in the social shaping tradition. Sociotechnical spaces may indeed harbour active elements like engineering practices, design approaches, project templates and management concepts which appear to play a key role in the (re-)configuration of a design or project space and the performance of its actors. In that sense, the tools researchers involve towards instigating reflexivity within transitions are not only concerned with transforming others' practices. Rather, at the same time through an ongoing responsiveness they are concerned with learning to learn alongside a diversity of peoples and researchers, in the production of engineering knowledge transforming their own actions as researchers.

As discussed in Chapter 2 (this volume), intermediary objects may play a significant role in the configuration of sociotechnical spaces. From a design anthropology perspective Vinck and Jeantet (1995) coin the notion of intermediary objects, pointing at the heterogeneous nature of intermediaries as networks of social actors and objects mediating between stages in design processes. Accordingly, whether the intermediary object will perform as a stable platform for the movement of knowledge across boundaries or whether it translates or mediates knowledge in a transformative way will depend on the stability of the objects, the social actors involved and their interrelations (Vinck and Jeantet 1995).

Building upon these methodological positionings, importance is given to how intermediary objects are staged within temporary collaborative research spaces and how they perform in practice. By taking intermediary objects as networks we intend to trace how what they perform depends on how they are becoming configured. The term 'temporary collaborative research space' refers to the delimitation and analysis of the participatory workshops discussed below. The workshops were set up and staged by the research team and included the use of ethnographic provocations (Buur and Sitorus 2007; Boer et al. 2013).

Our interest here is in what kinds of knowledge outcome are generated through purposely staged interactions between designers, design anthropologists and engineers in a temporary space mediating use and design, and whether we can trace the uptake of such knowledges into companies and organizations having a stake in designing indoor climate. How then is knowledge generated, packaged, transported and unpacked when moved from user domains to the 'home' organizations of the participating stakeholders?

RESEARCH METHOD

To illustrate the travel and/or mediation of user-oriented knowledge to instigate transitions we report on a case concerned with participatory innovation within the designing of indoor climate. The case study enabled us to trace the movement and transformation of knowledge on user practices from sites in Denmark and Germany to confront established configurations of users in worlds of institutionalized indoor climate research and development in line with Hyysalo and Johnsons' (2015) conception of 'the user' as a relational entity. The participatory exchange of knowledge took place in a temporary space including the SPIRE participatory innovation research team consisting of designers and design anthropologists and stakeholders such as engineers from an indoor climate research unit of a technical university, and engineers from companies developing and manufacturing components for indoor climate solutions such as windows, ventilation and control systems and insulation material etc.

Our report on the staging of the temporary space (in the format of three participatory workshops) draws on original accounts of the workshops including working documents, published conference papers as well as our own observations as participants in the workshops and six follow-up semi-structured interviews with project partners. Our interviews sought to trace end-user voices in the format of narratives of user practices from the participatory workshops into the realms of companies manufacturing building components and an established indoor climate research environment. Our aim was to analyse how the qualitative user knowledge was taken up, rejected or transformed in the participating organizations and their practices. In-depth 2:1 semi-structured interviews were carried out by the authors with six engineers from three companies from the Danish building sector and a university-based climate research centre – all being active in the SPIRE participatory innovation workshops. In the following detailed account of these interviews we mainly draw on two sites of uptake as illustration, the research institution and an industrial organization.

STAGING A TEMPORARY SPACE

The core idea of the 'Indoor Climate and Quality of Life' project was to 'take the perspective of the "user" – the occupants of homes, offices and institutions – rather than the usual position of the engineers who design, build and control indoor climate' (Buur 2012: 5). The project set out to investigate (a) what the notion of comfort in indoor climate means to people and how people experience comfort during their everyday use practices; and (b) what innovation potential rests in an appreciation of peoples' everyday practices and accounts of these practices. The project exemplifies the participatory innovation approach developed by the SPIRE centre, which 'seeks to combine the strengths of participatory design and design anthropology, while expanding towards a market orientation' (Buur and Matthews 2008: 268).

The staging of the temporary collaborative research space was carried out by the SPIRE team of designers and design anthropologists with a project manager and the head of the SPIRE centre (also taking on the role as a main facilitator) in a leading role. The selection of the participating engineers was due to a longer process of engaging companies and the research unit during the application for funding process seeking to include stakeholders to represent research and component manufacturers from the building sector. Regulatory bodies were not included but a number of the engineers were engaged in standardization work and building code negotiations.

The SPIRE research team organized a series of consecutive interactive stakeholder workshops and included activities such as sensemaking of field study material. Three of the workshops were dedicated to co-analysis of material (excerpts from interview transcripts with families, nursery teachers and office workers, video clips, photographic stills, etc.) from multi-sited field studies following indoor climate inhabitants and their practices carried out with people in their homes, nurseries and workplaces during 2009.

Engagement of engineers and researchers in collaborative sensemaking of field study material reflected a key navigational decision made by the SPIRE researchers to refrain from reliance upon established engineering concepts and understandings of comfort and indoor climate. The idea was to challenge the dominant engineering understandings of indoor climate and avoid an engineering focus on technological objects such as climate models and building components like insulation materials, ventilation installations, control equipment or windows (Buur 2012). The implication was that problem framings and design strategies based in the participating organizations were only indirectly included in the sensemaking process and not foregrounded in workshop interactions. Instead, throughout the workshops, the SPIRE team encouraged the participants to engage with the material to develop a notion of indoor climate that challenged their usual perspectives.

Methods for presenting and analysing empirical material varied across the workshops, from using excerpts from interviews and corresponding stills from video materials collated during field inquiries and involvement of performative tools for co-analysis. Co-analysis aimed towards analysing the framing of problems and solutions in relation to carefully selected user statements. According to an account of the workshops from members of the SPIRE team (Jaffari et al. 2011) a shared understanding of indoor climate practices gradually developed across workshop participants. An intermediate outcome of this shared understanding was manifested in the form of so-called 'comfort themes'. The 'comfort themes' included a package with six small booklets aimed towards moving into the spaces of uptake inhabited by the participating engineers. They provided the engineers with statements and selected field study material and user statements to work with in their own organizations to aid the generation of innovative ideas concerning indoor climate solutions.

During the workshops the facilitator, an accomplished design researcher with a background in engineering, pressed for the exploration of how these 'comfort themes' might translate into innovation potentials (Gunn and Clausen 2013). Here the facilitator presented a selected version of the comfort themes focusing on innovation potentials of building a relation with end-users of indoor climate during engineering design processes and practices. For example, in the booklet, 'Being Healthy' the authors asked: 'How would you enable people to act on and learn from their feelings of health within an indoor climate system of control?' Underpinning another theme 'Comfort is a political construct', the conception of indoor climate as a neutral standard based on 'what the experts say' was challenged. A number of the comfort themes presented caused controversial debate during the workshops bringing to the fore a clash between opinions of how user knowledge could be a source of innovation potential. The idea of enabling people to better control their indoor climate was problematic for one engineer: 'I see more trouble than innovation with that theme because the communication part is enormous' (Engineer from the skylight manufacturer).

INDOOR CLIMATE MODEL-BASED RESEARCH PRACTICES AND USER RELATIONS

Three of the engineers participating in the workshop series were affiliated with a Danish university-based research centre for indoor climate. The research centre for indoor climate has been an important player in the development of research-based indoor climate models and represents an internationally highly recognized research environment. The centre promotes itself as an important provider of knowledge to governmental regulation of building requirements as well as industry standards and practices in indoor climate.

During our interviews, the engineers from the research centre stressed how different the knowledge practices at the centre were from the local and qualitative knowledge they engaged with during the SPIRE workshops. Here their modelling practices stand out as a main reference point when comparing the differences between knowledge practices. The ambition of climate model research has typically been to describe the general relations between certain indoor climate factors and a measure of general satisfaction with the indoor climate based on a generalized 'user'.

Over the years, such models have provided an important frame of reference for building requirements and engineering standards in industrialized countries. These models have been criticized (Shove 2003; Jaffari and Matthews 2009) for framing a particular knowledge base, fuelling the development of an industrial indoor climate control regime based on uniform global industrial standards for indoor climate solutions independent of local climate and local cultures.

As one senior engineer from the university research centre said:

> Our group has always been thinking about humans in our research, and have involved people in our research by asking them how they perceived indoor climate ... But, even if we ask people what they feel, it is another question entirely what they actually do to control their environment. This opens up the question: Should we design a centrally controlled indoor climate environment, or should we delegate the control to people? ... This is an important topic in our research community, as it is now, it is a one size fits all, and it seems like there is an increasing tendency to challenge this.

While this statement points towards configuration of key social and political dimensions in modelling work, it is also clear that the tradition has a tendency to detach ideas of the social from engineering design practices (Harvey and Knox 2015) and even to remove any space in the models for certain kinds of qualitative knowledge. This has resulted in an understanding of the indoor climate user as being passive rather than active while negotiating indoor climate.

Similarly, one of the younger researchers from the same unit commented:

> For a long time, the climate models developed here were based on laboratory experiments with test persons in an artificial but controlled environment, but this did not represent real life situations. Recently, we have become more interested in human behaviour because you cannot explain or predict energy consumption in a house with natural ventilation without taking the human aspect into consideration.

There is a clear indication that a movement in knowledge practice would be possible based on a conscious linkage between the new energy saving agenda and a challenge of the current generalized user construction accentuated by the SPIRE project. Field studies may, according to these researchers, add new dimensions to understanding how people are using buildings and why they behave as they do, how they are affected by and how they influence indoor climate. Here it is important to remember emphasis is placed upon an understanding of human behaviour based on a behavioural psychology. Similar conceptions are reported from engineering practices concerned with users within the energy sector (Løgstrup et al. 2013).

While the younger research unit engineer experienced the SPIRE workshops as highly revealing and inspiring for future research, he also recalled the difficulties in following an ethnographic approach.

> With our engineering background we were actually not really able to interpret the video stills, video clips and narratives from field studies but were dependent upon the SPIRE researchers ... These experiences have opened up for some movements in our knowledge practices ... I have learned a lot and I will certainly take up the qualitative methods while conducting interviews as part of my future projects, as it gives a much better understanding of why people act as they do.

These university employed engineers' construction of an image of indoor climate users have been based on a search for factors or parameters depicting measurable effects on human well-being or on performance indicators. These laboratory settings form a certain kind of epistemic culture (Knorr Cetina 2001) allowing only certain framings of research questions. Often these practices, according to the younger researcher, assume an application of the results in specific engineering design practices based on single factor design requirements.

Current indoor climate modelling practices are furthermore an integral part of a larger engineering system with its infrastructure defining a chain of user constructions, indoor climate definitions and practices of usage in regulation, calculative methods and engineering design. This particular epistemic culture signifies the relevance of research as to whether it can be successfully translated into design recommendations in engineering practices and/or by defining a building design category. But the offering of clear-cut design recommendations also poses some dilemmas as indicated by one senior research unit engineer:

> The industry has been asking us to combine the diverse climate dimensions into one single measure, I don't know whether it is feasible or at all possible.

A younger research engineer working in the same unit supported this:

> Building engineers often expect a single figure in order to make design easier for themselves, but they often fail to understand the limitations and underlying complexity of indoor climate models.

Here, our young engineer refers to practice in certain engineering worlds, where engineers are reluctant to take on design responsibility and prefer to automate decisions or make rule-based decisions.

Temporary collaborative research spaces are not just any kind of temporary space or workshop conducted by a design consultancy. They are very specific. In this regard staging them must indicate what are the specific realities including ambiguity. We registered a movement away from the current and prevailing generalized understandings of end-users of indoor climate. But this movement was rather vague and constrained by established ideas of producing single dimensions and even a single figure as a design recommendation and the expectation of providing explanations and predictions of *user behaviour*. In this sense the user is still reduced to a variable in the engineering calculation. And the idea of seeing the user as a co-constructor/designer of indoor climate not to say a key player or subject in the control of indoor climate solutions can hardly be addressed within such a practice.

MARKETING STRATEGY CONFIRMATIONS AND USER RELATIONS

Similar difficulties in taking forward the distinctive perspective from the workshops can be seen in the context of inputs to product development and marketing. The engineer participating in the workshop from the skylight window manufacturer is located in a group of engineers and architects at the company headquarters. Their task is to provide technical marketing support to a global sales organization of a large (10 000 employees) Danish multinational company and to provide analysis and knowledge support to management decisions. The company is specialized in the development, manufacturing, marketing, and sales of one product – a skylight window – in a number of variants. This particular product has a strong position in a niche market globally and the company is the main branch of a leading Danish group in the building sector.

When the engineer from the skylight company compares the knowledge traditions of the SPIRE group and the company, he first of all refers to the unique company history and its organizational culture.

> This company has a very long tradition for quality and trustworthiness. Every statement from the company, therefore, has to be based on sound evidence. And here I mean based on technical arguments or on numbers … Only quantitatively based arguments are recognized as valid in top management and in the sales and marketing department. The same is the case when we want to present our point of view in standardization committees and revision of building regulations … We have a strong relation to research institutions like the Danish university-based research centre for indoor climate where it is important to be able to base your arguments in research-based data.

Many of the engineers in the company are recruited from the Danish research centre for indoor climate previously discussed and have similar engineering backgrounds and perceive arguments originating from this research as hard currency in political arguments for the building industry and its regulatory bodies. This observation underlines the existence of shared knowledge practices in an engineering world cutting across several organizational boundaries. Our engineer from the skylight engineering company expresses a strong awareness of these rules for making accounts in the organization. He was also aware, however, that the same rules were a strong barrier for the dissemination and sharing of knowledge from the SPIRE user-oriented project in the organization.

On the other hand, our engineer is also on the lookout to bring new approaches into the organization and tries to make an opening for anthropological knowledge by stressing the important role scientific institutions play in producing credibility in his organization. As he said:

> I noticed that the methods SPIRE used were also based on scientific arguments, and I like the idea that more university units contribute to research in indoor climate.

Actually, the company had recently employed an anthropologist to do user studies, which indicates that several knowledge practices could potentially co-exist in his organization, though the value of these accounts might vary according to circumstances.

The engineer from the skylight manufacturer clearly appreciated his engagement with qualitative knowledge, but, as he also points out, there seems to be a very limited uptake in the wider organization. Especially when it comes to the identification of innovation potential, the engineer has difficulties in pointing out where innovation takes place in the organization. In this instance, field-based material is most often confined to address, and to be appreciated in, the marketing processes and not in the front end of innovation. The company had a small front-end unit focusing on the development of new business areas, but this unit did not seem to pay much attention to findings from collective sensemaking generated within the SPIRE workshops.

TRANSITIONING ENGINEERING SYSTEMS?

Our interviews showed how engineers' unpacking of the 'comfort themes' within their 'home' organizations and engineering worlds related and reviewed received new understandings of user practices to established engineering knowledge practices. In particular, model-based assumptions of the end-user's role in the design of indoor climate and role end-users can (or could) play in designing indoor climate were challenged. But, while our engineers readily engaged in a deliberation of the possibilities for increasing engagement with end-users appreciating them as competent players with relevant knowledge, the engineers were rather hesitant about the possibilities of assigning users more active roles in the design of indoor climate solutions. The engineers we interviewed mainly related comfort themes to (enact) existing products, marketing, and business strategies and infrastructures of engineering models and systems. While this is not surprising, we observed a variety of patterns of uptake of qualitative understandings of user practices. In cases where user practices identified confirmed strategic concerns in the company or research unit, and where user knowledge was in line with dominant framings, as in product marketing, we found an unproblematized application of the 'new' user insights.

By reinterpreting the knowledge base of established user constructions, a number of current framings of users embedded in institutional structures and knowledge infrastructures and their associated taken for granted assumptions became challenged through the engineers' participation in the temporary space. Generally, we found the movements in understandings of users and use practices within the wider organizations to be rather limited. It seems like the participating engineers' own conceptions were highly challenged and moved, but their new insights clashed with established practices and made changes in engineering practices in the project partner organizations difficult if not impossible. This finding mirrors experiences with the governance of systems from the inside as discussed by Smith and Stirling (2007) according to which different actors participating in processes of transitioning systems often hold incommensurable framings of problems and solutions in relation to the reproduction of the sustainability of systems. It seems that it is possible – albeit no straightforward process – to challenge such dominant framings by questioning constructions of the user inherent in engineering models and practice.

TRAVEL OF INTERMEDIARY OBJECTS

How did the 'comfort themes' with their inscribed practice-oriented framing of indoor climate perform as intermediary objects moving from our temporary space to the established worlds of engineering (see Figure 7.1)? While the staged interaction between SPIRE researchers and stakeholders seems to perform a successful transformation of established user understandings within the temporary space, further attempts to reframe the relation between the private end-user and the stakeholder organizations operating in the worlds of indoor climate engineering seem much less successful. Similar observations stem from studies in other sectors and organizations even where concerns for user practices are articulated (Løgstrup et al. 2013; Brønnum and Clausen 2013).

User constructions (Akrich 1995) are abstractions that are constructed and appear as contestable terrain whereby actors from the diverse companies and

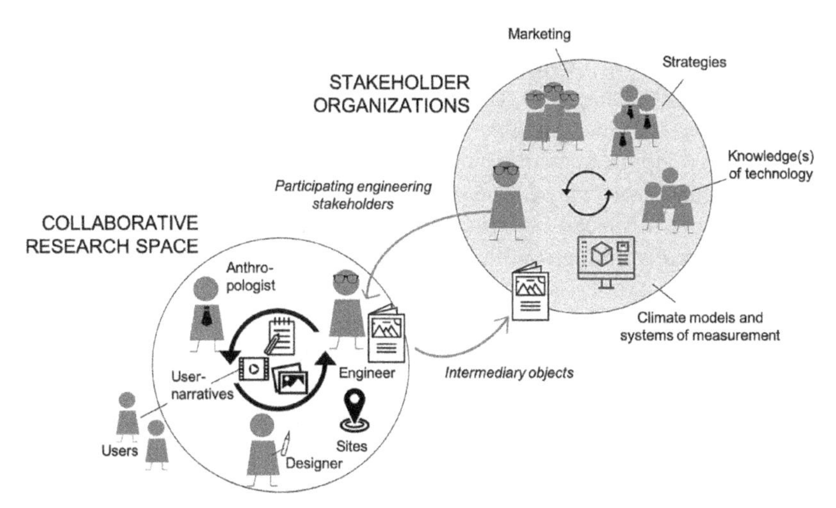

Note: Interactions in the temporary collaborative research space across the worlds of anthropologists, designers and engineers and the subsequent travel of intermediary objects from this space to stakeholder organizations.

Figure 7.1 Temporary collaborative research space

the indoor climate research institution seek to position themselves according to how they frame professional or strategic interests. As a consequence, the user constructions inscribed in the comfort themes became contested as they moved from the temporary space and into these engineering worlds. The 'comfort themes' can be viewed as heterogeneous intermediary objects (Vinck and Jeantet 1995) where the stability of the object depends on the staging and ongoing stabilization of relations between participants and objects engaged in the temporary space. While the participating engineers bring with them comfort themes from the temporary space into their partner organizations, SPIRE researchers and memories of others' everyday use practices become more distant while the engineering, organizational and research practices became evermore present.

TEMPORARY SPACES

Staging of temporary collaborative research spaces can hardly be expected to synthesize or transform engineering knowledge with a practice-based view of indoor climate use into full-blown innovative ideas for new products, systems or services. But, as in our case, they may include designing and framing intermediary objects, which moves forward ideas or conveys knowledge being able to travel to other design areas for further innovative iterations employing ingenuity and criticality. What we have seen in the SPIRE case is how a temporary space has problematized indoor climate conceptions and initiated a process towards raising awareness of the limitations of dominant knowledge infrastructures and constructions of users, which currently constrain and determine development activities in doing business. In this sense an outcome of this particular space may be the opening up of certain possibilities through the co-existence of several constructions of users and framing of design problems and alternative designs in the future compared to the single engineering model-based quest for certainties which have been the norm.

In line with Elgaard Jensen (2012) we demonstrate that dominant conceptions of the user have been challenged, but we cannot point at direct changes in design or knowledge practices in the participating companies. In many ways our findings echo studies of innovative work in larger mature companies (Dougherty 2008; Brønnum and Clausen 2013; Gish and Clausen 2013) where uptake of ideas from users to research and development (R&D) or knowledge transfer across knowledge domains is indeed difficult and demands a sustained effort over time, especially if these ideas are challenging taken for granted and entrenched knowledge practices.

A PROGRAMMATIC APPROACH

We have shown how a temporary space for participatory knowledge production was staged by a university research team as an attempt to enable transitioning of systems for indoor climate design and how it helped challenge and, under certain circumstances, even reframe existing engineering and model-oriented user conceptions towards a more relational understanding of user practice. But our study also shows that, due to strong path dependent innovation practices in the participating organizations, a direct uptake of a different understanding of user practices and their potential to reframe existing indoor climate engineering design practices beyond the temporary space proved limited.

A temporary collaborative research space as a sensitizing concept can contribute in exploring relations between use and design and the user as relational entity (Hyysalo and Johnson 2015: 84). By addressing the staging of temporary collaborative research spaces and intermediary objects, we have argued that intermediary objects such as 'comfort themes' can be subject to ongoing modifications in response to specific conditions and different groupings across different sites of indoor engineering practice. It appears that staging and configuration of a temporary collaborative research space and the design of intermediary objects are closely intertwined and mutually dependent. Importantly, such a temporary space whereby building relations between the movements of using and designing is made possible contribute towards making the production of indoor climate engineering knowledge more participatory.

Actionable Guidelines

Instigating a programmatic approach requires:

- criticality towards existing engineering models and decision-making processes, thereby enabling engineers to integrate multiple knowledge(s).
- willingness of social scientists to find ways of making their knowledges accessible to engineers towards building different kinds of quantitative indicators, sharing knowledges to address technical challenges, or improve technical specification, norms and standards. This may require social scientists learning to speak in technical norms and specifications. It is clear that this is not easy, and takes time and financial investment.

Adopting such an approach could:

- enable co-analysis of research materials while facilitating greater understanding across different disciplines and knowledge co-construction.

The case study in this chapter points to *challenging taken for granted assumptions*, which is a relevant strategy while adopting a programmatic approach. Limitation of gaining insights or understanding is often due to clashes with established practices. Thus, a way could be developed to:

- draw upon insights learned within temporary collaborative research spaces to inform future practices. This may mean greater attention to ingenuity as an important attribute of engineering practices (and associated objects).

'Comfort themes' became a relevant intermediary object in the temporary spaces discussed. Other intermediary objects, closer to new norms, specifications, indicators, models, technical challenges, problems to solve, new procedures or practice(s) and new organizational routines could be expanded further. Our findings show it is difficult to design outside an existing situation. For involved people in participatory innovation processes and practices, what would be the relevant intermediary objects?

Conducting research as a collaborative process attending to attunement of peoples' and technologies' everyday life histories it is important to:[4]

- raise awareness of the limitations of current knowledge infrastructures and constructions of users, which currently constrain and determine development activities in doing business.

- problematize various constructions of users and framing of design problems towards alternative ways of designing.

Recommendations

In the workshops outlined in this chapter, it appears that staging involves only a few social worlds, such as engineers within research and in product design, and users. Comfort is here taken as an example of enabling action: How to enable people to act on and learn from their feelings within the system of control? Comfort can be taken as a political construct, involving the same grouping of researchers and engineers, and traces of users. A programmatic approach could be expanded to find ways of engaging other actors involved in the building sector: companies (products, installations, service, maintenance), policy makers, architects, building managers, technicians who regulate control systems, etc. Regarding indoor climate engineers' hesitation about the possibilities of assigning users more active roles (in control of the system and/or in design), more data is required regarding what users do when they are excluded from the systems, for instance, deploying subversive practices and social relations. Research partners should design towards shifting imposition of energy strategies, so end-users can have more ownership and understanding of systems of measurement: how to negotiate with vested interests; greater involvement of government in supporting manufacturers concerned with improving indoor climate and quality of life.

NOTES

1. This chapter is a revised version of Clausen and Gunn (2015).
2. The TempoS project (2010–14) was funded by the Danish Strategic Research Council.
3. SPIRE (Sønderborg Participatory Innovation Research Centre) was a research centre funded by the Danish Government as part of the User Driven Development Programme (2008–13) and the Danish Strategic Research Council.
4. SPIRE contributed to the 'Designing Environments for Life Programme', funded by the Scottish Institute For Advanced Studies, hosted by the University of Strathclyde, September–December 2009. A number of multidisciplinary workshops were conducted with international academics, architects, designers and policy makers. The task of the programme was to bridge the gap between the familiar environments of quotidian experience and the environment of scientific, technological and policy discourse, in such a way that we are better able to reconnect the practices of everyday life with the imperatives of environmental sustainability.

REFERENCES

Akrich, M. (1995), 'User representations: Practices, methods and sociology', in A. Rip, J. Misa and J. Schot (eds), *Managing Technology in Society: The Approach of Constructive Technology Assessment,* London: Pinter, pp. 167–84.

Berkhout, F., A. Smith and A. Stirling (2004), 'Socio-technical regimes and transition contexts', in B. Elzen, F.W. Geels and K. Green (eds), *System Innovation and the Transition to Sustainability*, Cheltenham, UK and Northampton, MA, USA: Edward Elgar Publishing, pp. 48–75.

Binder, T. and E. Brandt (2008), 'The Design Lab as platform in participatory design research', *CoDesign*, **4** (2), 115–29.

Boer, L., J. Donovan and J. Buur (2013), 'Challenging industry conceptions with provotypes', *CoDesign*, **9** (2), 73–89, DOI: 10.1080/15710882.2013.788193.

Brønnum, L. and C. Clausen (2013), 'Configuring the development space for conceptualization', in U. Lindemann et al. (eds), *Proceedings of the 19th International Conference on Engineering Design (ICED13) Design for Harmonies*, Seoul: Design Society, pp. 171–80.

Buur, J. (ed.) (2012), *Making Indoor Climate: Enabling People's Comfort Practices*, Sønderborg: Mads Clausen Institute, University of Southern Denmark.

Buur, J. and B. Matthews (2008), 'Participatory innovation', *International Journal of Innovation Management*, **12** (3), 255–73.

Buur, J. and L. Sitorus (2007), 'Ethnography as design provocation', in *Proceedings of Ethnographic Praxis in Industry Conference (EPIC 2007)*, Hoboken, NJ: John Wiley and Sons, pp. 140–50.

Clausen, C. and W. Gunn (2015), 'From the social shaping of technology to the staging of temporary spaces of innovation: A case of participatory innovation', *Science & Technology Studies*, **28** (1), 73–94.

Clausen, C. and C. Koch (2002), 'Spaces and occasions in the social shaping of information technologies', in K.H. Sørensen and R. Williams (eds), *Shaping Technology, Guiding Policy: Concepts, Spaces and Tools*, Cheltenham, UK and Northampton, MA, USA: Edward Elgar Publishing, pp. 223–48.

Clausen, C. and Y. Yoshinaka (2005), 'Socio-technical spaces – guiding politics, staging design', *International Journal of Technology and Human Interaction,* **2** (3), 44–59.

Clausen, C. and Y. Yoshinaka (2007), 'Staging sociotechnical spaces: Translating across boundaries in design', *Design Research*, **6** (1–2), 61–78.

Dougherty, D. (2008), 'Managing the "unmanageables" of sustained product innovation', in S. Shane (ed.), *Handbook of Technology and Innovation Management*, Hoboken: John Wiley and Sons, pp. 173–93.

Elgaard Jensen, T. (2012), 'Intervention by invitation: New concerns and new versions of the user in STS', *Science Studies*, **25** (1), 13–36.

Gaziulusoy, A. and H. Brezet (2015), 'Design for system innovations and transitions: A conceptual framework integrating insights from sustainability science and theories of system innovations and transitions', *Journal of Cleaner Production,* **108**, 558–68.

Geels, F.W. (2002), 'Technological transitions as evolutionary reconfiguration processes: A multi-level perspective and a case-study', *Research Policy*, **31**, 1257–74.

Gish, L. and C. Clausen (2013), 'The framing of product ideas in the making: A case study on the development of an energy saving pump', *Technology Analysis and Strategic Management*, **25** (9), 1085–101.

Gunn, W. and C. Clausen (2013), 'Conceptions of innovation and practice(s): Designing indoor climate', in: W. Gunn, T. Otto and R.C. Smith (eds), *Design Anthropology: Theory and Practice*, London: Bloomsbury, pp. 159–79.

Harvey, P. and H. Knox (2015), 'Virtuous detachments in engineering practice – on the ethics of (not) making a difference', in T. Yarrow, M. Candea and J. Cook (eds), *Detachment: Essays on the Limits of Relational Thinking*, Manchester: Manchester University Press, pp. 58–78.

Hyysalo, S. and M. Johnson (2015), 'The user as relational entity', *Information Technology & People*, **28** (1), 72–89.

Irwin, T., G. Kossoff, C. Tonkinwise and P. Scupelli (2015), *Transition Design 2015*, Pittsburgh, PA: Carnegie Mellon University.

Jaffari, S. and B. Matthews (2009), 'From occupying to inhabiting – a change in conceptualising comfort', *IOP Conference Series: Earth and Environmental Science*, **8** (1), 1–14.

Jaffari, S., L. Boer and J. Buur (2011), 'Actionable ethnography in participatory innovation: A case study', in *Proceedings of The 15th World Multi-conference on Systemics, Cybernetics and Informatics, Vol. 3.* Orlando FL: International Institute of Informatics and Systemics, pp. 100–6.

Knorr Cetina, K. (2001), 'Objectual practice', in T.R. Schatzki, K. Knorr Cetina and E. von Savigny (eds), *The Practice Turn in Contemporary Theory*, London: Routledge, pp. 175–88.

Løgstrup, L.B., M.M. Nelson, W.S. Mosleh and W. Gunn (2013), 'Designing anthropological reflection within an energy company', in *Proceedings of Ethnographic Praxis in Industry Conference, EPIC 2013, 15th–18th September, London*, Hoboken, NJ: John Wiley and Sons, pp. 116–28.

Shove, E. (2003), *Comfort, Cleanliness and Convenience: The Social Organization of Normality*, Oxford: Berg.

Smith, A. and A. Stirling (2007), 'Moving outside or inside? Objectification and reflexivity in the governance of socio-technical systems', *Journal of Environmental Policy & Planning*, **9** (3–4), 351–373.

Tonkinwise, C. (2015), 'Design for transitions—from and to what?', *Design Philosophy Papers*, **13** (1), 85–92.

Tonkinwise, C. (2017), 'Post-normal design research: The role of practice-based research in the era of neoliberal risk', in L.Vaughan (ed.), *Practice Based Design Research*, London: Bloomsbury, pp. 29–39.

Vinck, D. and A. Jeantet (1995), 'Mediating and commissioning objects in the sociotechnical process of product design: A conceptual approach', in D. Maclean, P. Saviotti and D. Vinck (eds), *Designs, Networks and Strategies, Vol. 2, COST A3 Social Sciences*, Brussels: EC Directorate General Science R&D, pp. 111–29.

8. Staging a circular economy journey

Rikke Dorothea Huulgaard, Eva Guldmann and Søren Kerndrup

INTRODUCTION

In recent years, the circular economy (CE) has been promoted as an economic paradigm that can provide the necessary leverage towards sustainable development. The CE has been well described on a conceptual level (Ellen MacArthur Foundation 2013; Ghisellini et al. 2016), but less so as an innovation process—that is the process of finding circular solutions that are relevant to the individual company.

The transition towards a CE demands a shift from current approaches to establish new production, consumption and transportation patterns and develop new energy systems to support them. Such a shift creates new opportunities for development and innovation, and at the same time threatens to disrupt existing business and operating practices and thus challenges current lock-ins at the mental, technological, economic, organizational and institutional levels (Guldmann and Huulgaard 2020; Doganova and Karnøe 2012). In response, scholars are beginning to examine how to help companies make such transitions (Geissdoerfer et al. 2016, 2017a, 2017b; Pedersen and Clausen 2018). In particular, they are beginning to explore the importance of experimentation as a means to overcome inertia (Weissbrod and Bocken 2017), create awareness, and develop new solutions and networks. The existing literature focuses primarily on the development and use of specific tools to promote experimentation (Bocken et al. 2018); overall, there has been less emphasis on the importance of the contexts in which these tools are applied. We argue that the deliberate design of the contextual setting for experimentation—defining who to involve, where and when, and so forth—has to be considered in tandem with the tools themselves, and we conceptualize the contextual design as an instance of staging a temporary space to stimulate experimentation and interaction among different stakeholders (Pedersen and Clausen 2018).

The shifts that are required for companies to transition from a linear economy to a CE render staging especially important for incumbent companies, which are embedded in existing value chains and business models; such companies can miss opportunities that arise from a transition to a CE due to an inability to overcome inertia and lock-ins (Guldmann and Huulgaard 2019). In this chapter, we focus on how temporary spaces can be staged to inspire businesses to search for and experiment with activities to facilitate the transition to a CE. Specifically, we ask: How can the staging of temporary spaces contribute to the transition from a business logic based on a linear economy paradigm to a logic based on a CE paradigm? We examine this question via an in-depth case study of how an incumbent company recently embarked on a journey to explore potentials in a transition to CE. In the next section, we introduce the main theoretical concepts that have framed our understanding of staging temporary spaces. Then, we describe our research design and provide background information about the case company. We describe three temporary spaces staged at the case company, and analyse impacts of the different space configurations before offering some concluding remarks.

THEORETICAL BACKGROUND

Although CE is born out of a long theoretical and practical tradition of among others the self-replenishing system (Stahel 1982), ecodesign (Brezet et al. 1997), green technology (Remmen 2001), and as such many elements in the CE paradigm are not new, the key difference between the CE paradigm and earlier sustainability paradigms is the ensuing changes to a company's business models. CE focuses on keeping products, components and materials in circulation for as long as possible. This sets new requirements for the products, for instance that they are upgradeable, reparable and modular. It also requires redesign of the way that the product is brought to the customers—involving changes to the value proposition, the customer relationships and possibly distribution channels—to ensure the product is returned to the original equipment manufacturer (OEM) or a third party for reuse, refurbishment or recycling. Implementing a circular business model will therefore affect most parts of an existing business model including the cost and profit structure (Guldmann et al. 2019). This kind of change is demanding, risky and traditionally not something companies would engage in unless they face business-critical situations like eroding margins or increasing competitive pressure that threaten profitability. However, identifying links between the CE and earlier sustainability paradigms, and between CE practices and existing or earlier practices in different parts of the company can provide a basis for turning such a challenge into an opportunity (Weich and Quinn 1999); and bridging between the new and the old in this way was something we, as researchers, actively worked to facilitate in the collaboration with the case company.

In this chapter, we focus on the challenges that lie in the first steps of a company's innovation journey, where actors explore how transitioning to a CE paradigm could make sense for the company. We are especially interested in companies that are not experiencing a burning platform (Zurn and Mulligan 2014) that can motivate change to the business model. In these companies, the challenge of transitioning from a linear production paradigm to a CE paradigm is about clarifying how and where it would be possible and beneficial to initiate and develop circular ideas, activities and routines in a context where operations, product and business development processes are dominated by a linear production paradigm embedded in dominant constellations of actors (Bucher and Langley 2016; Guldmann and Huulgaard 2019). An important step in any transition is sensemaking—in this case, with regards to the relevance of the CE and a key question is how to open up sensemaking processes that make CE-related activities meaningful in a situation where the company's dominant paradigm is a linear production paradigm (Weick et al. 2005). Developing new routines and sensemaking happens in a complex flow of activities and events within and outside companies; thus, it is important to examine how actors make sense of ongoing events (that are not actively planned or staged by actors in the company) by bracketing and linking them with planned events (that *are* actively staged) (Bojovic et al. 2020; Howard-Grenville et al. 2011). Hence, navigation and negotiation are important factors in the design of the staging process and thus parameters that are important to consider to successfully facilitate the transition to a business logic based on the CE.

Activities and sensemaking processes that are not hampered by the existing linear paradigm occur in various types of spaces, which can be understood as 'bounded social settings' that are socially, materially, temporally and symbolically separated from the logic, routines and activities of the dominant paradigm (Bucher and Langley 2016). The spaces that provide the framework for staging new activities and forming opinions may be planned or unplanned. Planned spaces are negotiated and designed to create bounded social settings that enable ideas to interact and activities to develop that are not embedded in the dominant logic and routines, for example, in the form of workshops and hackathons (Brandt et al. 2013; Flores et al. 2018). Unplanned spaces form when situations arise during everyday business activities that enable actors to challenge current actions and opinions and form new ones, for instance, when writing memos or making presentations at international conferences (Bucher and Langley 2016). It is therefore necessary that negotiations on staging of planned spaces take into account the developments in the internal and external unplanned spaces, as these can open up and close down opportunities—that is, the planned spaces are situated in relation to practices in and around the company (Orlikowski 2002). The needed attention to the ever-changing opportunity space places great demands on those who negotiate staging of the planned spaces, since they should utilize and design spaces in a way that leverages emerging opportunities.

One can deliberately design or stage a temporary space (Chapter 2, this volume). Such staging of a temporary space concerns how the space—that is the specific event—is prepared and arranged (Brodersen et al. 2008). It is about the sociomaterial constellation of the space, the goals, how the interaction is planned and who is included or excluded in the space, along with which objects are part of the negotiations (Brandt et al. 2005; Bucher and Langley 2016; Iversen and Leong 2012). Furthermore, staging is about how different actors, objects and meaning can be put together in a space in order to facilitate or enable interaction and co-creation (Chapter 2, this volume). In this way, the temporary spaces are balancing between two worlds—that is existing activities and understandings on one hand and possible future activities on the other (Orlikowski 2002). The stagers of the temporary spaces are facilitators of these negotiations rather than experts themselves (Pedersen 2020; Chapter 2, this volume). Hence, a gradual paradigm shift is not just a result of the stager's reasoning, but also of a collective interpretation and a translation that happens through negotiation in and between the temporary spaces (Pedersen 2020; Chapter 2, this volume).

Outcomes from such spaces depend on the extent to which it is possible to decouple creative processes from the dominant paradigm, and not least on how the creative processes are staged. Negotiations specifically on the staging of planned spaces can consequently be seen as an organizational change process (Dawson 2000), which is shaped by both internal and external political activities, and, hence, the context in which the change process takes place is central for the unfolding of the process and its outcome. Thus, how to stage is not just a product of the staging initiator's cognitive reflections but also influenced by the overall opportunity space, which is determined by the internal and external context and the ongoing negotiations between actors, which in our case included both negotiations between company actors and negotiations between company actors and researchers.

In summary, breaking with existing paradigms is a prerequisite for developing new ideas and establishing new interactions and activities, and it places great demands on actors' abilities to navigate a complex flow of activities. Hence, in order to succeed, it is important to create spaces (bounded settings), where it is possible to experiment with activities, routines and ways of thinking and at the same time to connect these activities to the existing activities, events and logics in ways that make them meaningful and significant for future development. In this chapter, we examine how actors, researchers and companies can collaboratively negotiate and stage temporary spaces that can form the basis for a transition towards a CE in a company dominated by a linear production paradigm.

METHODOLOGY

Data collected during close collaboration with the company from 2013 to 2017 form the empirical foundation for this chapter. The longitudinal research design enabled the researchers and company employees to build relationships based on trust and the researchers participated in negotiations to stage activities and spaces that facilitated and supported the transition to the CE over time.

Company Description

The focus of our case study is a large multinational company based in Denmark with more than 26,000 employees worldwide and four divisions, exemplifying a large incumbent company with complex organizational structures. Activities such as developing sustainability strategies and guidelines were handled primarily by a corporate function, whereas the operations-oriented divisions were responsible for implementing sustainability initiatives. At the beginning of our collaboration, the company was unfamiliar with the CE concept, but it was willing to learn more and evaluate its relevance.

The company activities were based on linear business models when the research collaboration began, supplemented by a sustainability focus on compliance, CSR and energy efficient products and a burgeoning interest in resource efficiency. This set up was reflected in existing mindsets and organizational procedures. As such, transitioning towards circular business models that substituted the existing business logic of one-time sale of energy efficient products with a logic of taking responsibility for and creating value from multiple product lifecycles would clearly constitute a paradigm shift for the company. The company had no urgent reason to implement such a drastic change to its business models and sustainability practices and, hence, the company constituted a good case to study how staging a series of temporary spaces can facilitate the transition towards CE in companies that experience little internal or external pressure to embark on such a journey.

Research Design and Data Sources

A number of events took place between 2013 and 2017 which influenced the unfolding of the CE journey that we analyse in this chapter, such as internal meetings and external seminars. However, in this chapter, we home in on three specific temporary spaces that were staged during that period. Throughout the study, we, the researchers, actively worked to organize temporary spaces and to monitor spaces initiated by other actors. We were also in close dialogue with key company contacts to track changes among actors and the company's overall progress as it embarked on its journey. The three temporary spaces differed in terms of overall purpose, the actors involved and configuration—that is, how they were organized and the interaction was planned, and which objects were utilized. The data sources for our analyses of the temporary spaces are interviews with key informants involved in developing the ecodesign guideline, direct observation of actors during the second and third temporary space, captured in field notes, official workshop minutes, follow-up interviews with four key informants to identify changes in beliefs and actions following the second and third temporary space. All interviews were recorded, transcribed and coded. Quotations from the interviews used in the case study have been translated from Danish by the authors.

Figure 8.1 illustrates the three temporary spaces (blue circles) and negotiations taking place between the researchers and the sustainability director in between the three temporary spaces. Figure 8.1 also exemplifies the external events that influenced the process. In the sections that follow, we analyse each of the temporary spaces that were staged at the company.

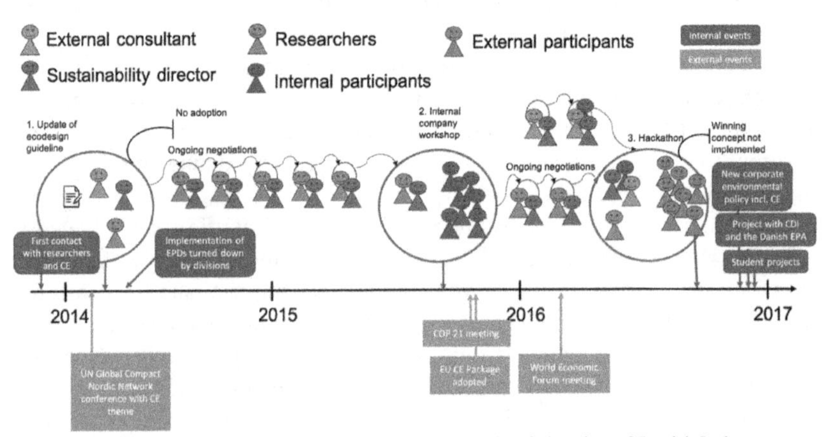

Note: EPA: Environmental Protection Agency; CDI: The Confederation of Danish Industry.

Figure 8.1 *Temporary spaces and internal and external activities: the three temporary spaces and the internal and external activities that influenced the process*

STAGING OF TEMPORARY SPACES

We first reached out to the case company concerning participation in our research project in late 2013 and our point of contact was the sustainability department. The decisive factor in support of the collaboration seemed to be the fact that circular business models hold the potential to increase resource efficiency, among other benefits, and this particular aspect aligned with the existing dominant sustainability paradigm.

Temporary Space 1: Expanding the Ecodesign Guideline to Include CE Principles

At the onset of the research collaboration we negotiated with the sustainability director around what spaces might be created to kick-off the exploration of the CE paradigm and the related new business logic. Heavy day-to-day workloads in the operational divisions of the company at the time meant that under these contextual conditions (Dawson 2000) the sustainability director preferred to begin the exploration at a smaller scale that involved only few actors. It was therefore decided to focus on a project that was already under way, and we were invited into a space delimited by the company: the revision of an ecodesign guideline. The ecodesign guideline is a mandatory document used in the product development process that provides development teams with a guide to consider various environmental impacts of the new product under development, and it was scheduled for an update to reflect the latest environmental regulation.

As such, the initial negotiations in the specific company context meant that we took advantage of an unplanned space that opened up because of other ongoing activities in a contextual setting where there was little room for our initial aim of involving more actors. The result of this space was that the guideline was not only updated to reflect new regulation, but expanded with advice to reflect on the possibility of using recycled or recyclable materials, dismantling the product into pure material fractions, reusing products, etc.

Although it may seem that the space had little impact, it did give the sustainability director a chance to get acquainted with the CE paradigm and to start reflecting on its implications. As such, this may have been an important factor in the decision of the sustainability director to stage a new temporary space that would involve more actors from across the organization.

Temporary Space 2: A Cross-organizational Workshop on the CE

Around 18 months passed until it was possible to stage another temporary space. During this time, CE activities were not a high priority and the company was hesitant to engage in any environmental improvements, for example, an internal project focused on implementing environmental product declarations (EPDs) (see Figure 8.1) was abolished because no division was willing to invest the needed resources in the project. Nevertheless, the researchers participated in negotiations via meetings, emails and phone calls with the sustainability director, who on his part conducted internal negotiations. The sustainability director and the researchers collectively succeeded in staging the second temporary space; a cross-organizational workshop in the autumn of 2015.

The aim of the temporary space was to introduce the actors to CE principles and generate ideas for relevant circular business models. The workshop involved 14 actors from two different divisions who held positions related to research and development (R&D), product management, quality assurance, industry affairs, regulatory compliance and sustainability. Four researchers, including two of the authors, also participated in the workshop. Several objects such as a large poster and post-it notes for brainstorming purposes were applied to facilitate sensemaking and spur a change in the existing sustainability paradigm.

Compared to the first temporary space, the effect of the second temporary space was larger, as it instigated learning among the actors that enabled them to make sense of CE and its relevance to the company. During the workshop, we observed how employees from various departments were able to link the examples offered by the researchers on CE principles to business activities that were a successful part of the business model of the case company many years ago. Actors proposed several ideas for how the divisions could benefit from introducing CE activities and how to combine these with existing knowledge and goals, such as by incorporating modularity into product design and service contracts. The workshop also highlighted challenges around the implementation of circular business models, such as institutional lock-ins related to regulation as well as organizational lock-ins related to difficulties in communicating the benefits of the CE in standard business case templates and needing management's approval prior to initiating small-scale pilot projects.

This temporary space seemed to have had a positive influence on attitudes towards CE, and in the months following the workshop, many external events and activities constituted unplanned temporary spaces that reinforced the company's willingness to experiment further (see Figure 8.1). For instance, top management attended COP 21 (the United Nations Paris Climate Conference) at the end of 2015 and the World Economic Forum meeting in early 2016, where CE was on the agenda. In preparation for these meetings, memos on CE were prepared for the CEO. Furthermore, the European Union (EU) adopted a CE policy in December 2015 (European Commission 2015). Consequently, the company's public affairs employees also had the CE on their agenda, and were trying to decipher what EU's CE package would imply for the company and to influence the debate in the EU.

The staging of the cross-organizational workshop successfully created a temporary space for the actors to explore and learn about the CE, and began to establish a common understanding and vocabulary amongst the actors.

Establishing a common language and visualizing ideas also triggered informal activities across the company's divisions. A hardware director explained: 'A network was generated at the [workshop]. I had never met [the sustainability director] before … Well it happens like these things happen, doesn't it? People meet and begin to talk … And then it grows from there. So [the workshop] has had that effect' (Interview 21 September 2016).

The second temporary space was also important because it attracted attention and spurred interest in a third temporary space: 'There is no doubt that the entire project that [the researchers] kick-started with the workshop has been a significant factor in making [the third temporary space] happen. I do not think that it would have happened if [the hardware director] had not participated in that workshop' (Sustainability director, interview 23 September 2016).

Temporary Space 3: CE Hackathon

The third temporary space was a CE hackathon that was proposed to the case company by an external consultancy. The hackathon focused on redesigning packaging for a product that was itself under redesign at the time. The aim was to find an alternative to the existing styropor packaging material.

Students from a nearby university constituted the main actors in the hackathon, while actors from the company included a specialist from one of the divisions and the sustainability director. The specialist presented the challenge to the students when the hackathon was kicked off and evaluated the various suggestions together with the sustainability director and a few others, when the hackathon was over, to decide on a winning design that would be presented to the packaging design team later on.

The participants were divided into three groups; each had 24 hours to develop a solution. The participants were given different presentations to inspire their work, including a presentation about CE principles. They also had materials and tools at their disposal to build prototypes.

The direct influence of this third temporary space was limited to providing inspiration for the case company: 'It is an obvious opportunity to get some external input, because sometimes it is also about getting external people to look at your problems with new eyes and new ideas' (Hardware director, interview 21 September 2016). The hackathon provided inspiration specifically around the use of alternative packaging materials such as moulded fibre packaging; however, the division that wanted ideas for a redesign of the specific packaging was already working on different moulded fibre packaging solutions, and the winning packaging design was not implemented.

DISCUSSION AND CONCLUSION

Our aim in this chapter was to analyse how to stage temporary spaces that promote a transition from a linear production paradigm to a CE paradigm, which demands a change both in business logic and sustainability paradigm. Our analysis showed how three temporary spaces contributed in varying degrees to develop the awareness of and interest in CE that is needed to support a beginning transition to a CE paradigm.

The experiences from the three temporary spaces contributed to our understanding of spaces as situated in a specific context (Orlikowski 2002) suggesting that staging is about creating spaces that are adapted and designed in relation to the internal and external company contexts that they are part of. This implies in part that the negotiations on how to stage temporary spaces is a political process where, in this case, the researchers and the sustainability director had to negotiate the conditions for staging the temporary spaces internally in the organization and adapt to the existing paradigm and to possibilities that arose. It also implies that external events (that is, unplanned spaces), where actors engage in activities that are outside the dominant business logic and sustainability paradigm are an integral part of the overall CE journey. For example, shortly after the second temporary space had been staged, top management from the company attended international meetings about CE and the EU adopted a new CE policy (see Figure 8.1) that meant the external context (Dawson 2000) changed. This in turn affected the internal context, which was observed as an increase in the divisions' interest in further CE experimentation. More specifically, company actors prepared presentations on CE for the top management as preparation for these external events, and later the company adopted a new corporate environmental policy including CE (see Figure 8.1).

Our analysis also illustrated how the outcome of the temporary spaces depended on how the spaces were configured and who was invited. The first temporary space concerned with the revision of an ecodesign guideline and the third space in the form of a hackathon had limited impact in the company, which seemed to link to the fact that external actors were the main contributors in these spaces. In contrast, the second space was an experimental space delimited from daily operations. Here the potential for changing the existing sustainability paradigm and anchoring thereof was relatively large due to the involvement of actors from different departments, who together were given the opportunity to experiment and make sense of how CE could be relevant to the company (Weick et al. 2005).

The stagers' navigation and negotiation play an important role in staging temporary spaces (Pedersen 2020). If navigation of the context and the political process of configuration of the temporary spaces are not successful, the staged spaces may not yield the wanted outcomes. The hackathon is a case in point: the choice of challenge to work on, the few company actors involved and the lack of strong CE guidance for the external participants resulted in a discouraging outcome at a phase in the innovation journey where more prominent outcomes should be possible. The need to navigate and negotiate depending on the context requires a sensitivity in relation to the concrete situation of the company and its groups of actors on the stagers' part. This means it is important to work closely with company actors to understand the company context—that is, strategies, priorities, workloads, power dynamics, history, etc. as well as different ways of seeing and doing things—which will offer guidance as to what actors could be attracted, what configurations of spaces are possible and what are realistic outcomes of the spaces.

In conclusion, the three temporary spaces successfully began the journey to unlock inertia, but they each played a different role and were not equally impactful. The key to successfully stage the CE journey is to leverage the openings that arise for innovation and sensemaking in the uncertain and ever-changing field of opportunities and limitations, which is a particular form of flexible and emergent staging. The aim is to take advantage of emergent openings to create temporary spaces where it is possible for actors to experiment in ways that allow activities to be developed outside the dominant routines and tasks and do so in a way that allows for subsequent shifts in the company's sustainability paradigm.

When introducing new paradigms that have the potential to disrupt existing business logics, it is important to be open towards and navigate the changing opportunity field, which is affected by internal and external contexts and political processes. When staging temporary spaces it is therefore imperative to assess the company internal and external context and negotiate the best possible configuration and sequence of temporary spaces for working with the new paradigm. Finally, it is important to work closely with company actors to understand the company context (strategies, priorities, workloads, power dynamics, history, etc.), which will offer guidance as to what staging is possible, that is, what actors could be attracted, what configurations of spaces are possible and what are realistic outcomes of the spaces.

ACKNOWLEDGEMENTS

The authors would like to thank the employees from the company that participated in the research project and in interviews for permitting us to influence and follow their CE journey. This work was supported by the Danish Environmental Protection Agency and the Danish Industry Foundation.

REFERENCES

Bocken, N.M.P., C.S.C. Schuit and C. Kraaijenhagen (2018), 'Experimenting with a circular business model: Lessons from eight cases', *Environmental Innovation and Societal Transitions*, **28**, 79–95.

Bojovic, N., V. Sabatier and E. Coblence (2020), 'Becoming through doing: How experimental spaces enable organizational identity work', *Strategic Organization*, **18** (1), 20–49.

Brandt, E., T. Binder and E. Sanders (2013), 'Tools and techniques: Ways to engage telling, making and enacting', in J. Simonsen and T. Robertson (eds), *Routledge International Handbook of Participatory Design*, New York: Routledge, pp. 145–81.

Brandt, E., M. Johansson and J. Messeter (2005), 'The design lab: Re-thinking what to design and how to design', in T. Binder and M. Hellström (eds), *Design Spaces*, Finland: Edita Publishing Ltd IT Press, pp. 34–43.

Brezet, H., C. van Hemel and R. Clarke (1997), *ECODESIGN: A Promising Approach to Sustainable Production and Consumption*, Paris: United Nations Environment Programme (UNEP).

Brodersen, C., C. Dindler and O.S. Iversen (2008), 'Staging imaginative places for participatory prototyping', *CoDesign: Design Participation(-S)*, **4** (1), 19–30.

Bucher, A. and S. Langley (2016), 'The interplay of reflective and experimental spaces in interrupting and reorienting routine dynamics', *Organization Science*, **27** (3), 594–613.

Dawson, P. (2000), 'Technology, work restructuring and the orchestration of a rational narrative in the pursuit of "management objectives": The political process of plant-level change', *Technology Analysis & Strategic Management*, **12** (1), 39–58.

Doganova, L. and P. Karnøe (2012), *The Innovator's Struggle to Assemble Environmental Concerns to Economic Worth: Report to Grundfos New Business, March 2012*, Bjerringbro: Grundfos New Business, accessed 5 August 2020 at https://www.industriensfond.dk/sites/ default/files/grundfos.pdf.

Ellen MacArthur Foundation (2013), *Towards the Circular Economy: Economic and Business Rationale for an Accelerated Transition*, Vol. 1, UK, accessed 5 August 2020 at https://www.ellenmacarthurfoundation.org/assets/downloads/publications/-Ellen-MacArthur-Foundation-Towards-the-Circular-Economy-vol.1.pdf.

European Commission (2015), *Closing the Loop – An EU Action Plan for the Circular Economy*, Brussels: European Commission.

Flores, M., M. Golob, D. Maklin, M. Herrera, C. Tucci, A. Al-Ashaab, L. Williams, A. Encinas, V. Martinez, M. Zaki, L. Sosa and K.F. Pineda (2018), *How Can Hackathons Accelerate Corporate Innovation?*, Cambridge: University of Cambridge.

Geissdoerfer, M., N.M.P. Bocken and E.J. Hultink (2016), 'Design thinking to enhance the sustainable business modelling process—A workshop based on a value mapping process', *Journal of Cleaner Production*, **135**, 1218–32.

Geissdoerfer, M., P. Savaget, N.M.P. Bocken and E.J. Hultink (2017a), 'The circular economy—a new sustainability paradigm?', *Journal of Cleaner Production*, **143**, 757–68.

Geissdoerfer, M., P. Savaget and S. Evans (2017b), 'The Cambridge Business Model innovation process', *Procedia Manufacturing*, **8**, 262–9.

Ghisellini, P., C. Cialani and S. Ulgiati (2016), 'A review on circular economy: The expected transition to a balanced interplay of environmental and economic systems', *Journal of Cleaner Production*, **114**, 11–32.

Guldmann, E. and R.D. Huulgaard (2019), 'Circular business model innovation for sustainable development', in N. Bocken, P. Ritala, L. Albareda, and R. Verburg (eds), *Innovation for Sustainability: Business Transformations Towards a Better World*, Cham: Springer International Publishing, pp. 77–95.
Guldmann, E. and R.D. Huulgaard (2020), 'Barriers to circular business model innovation: A multiple-case study', *Journal of Cleaner Production*, **243**, accessed 5 August 2020 at https://doi.org/10.1016/j.jclepro.2019.118160.
Guldmann, E., N.M.P. Bocken and H. Brezet (2019), 'A design thinking framework for circular business model innovation', *Journal of Business Models*, **7** (1), 39–70.
Howard-Grenville, J., K. Golden-Biddle, J. Irwin and J. Mao (2011), 'Liminality as cultural process for cultural change', *Organization Science*, **22** (2), 522–39.
Iversen, O.S. and T.W. Leong (2012), 'Values-led participatory design: Mediating the emergence of values', in *Proceedings of the 7th Nordic Conference on Human-Computer Interaction: Making Sense Through Design*, New York, NY: Association for Computing Machinery, pp. 468–77.
Orlikowski, W.J. (2002), 'Knowing in practice: Enacting a collective capability in distributed organizing', *Organization Science*, **13** (3), 249–73.
Pedersen, S. (2020), 'Staging negotiation spaces: A co-design framework', *Design Studies*, **68**, 58–81.
Pedersen, S. and C. Clausen (2018), 'Co-designing for a circular economy', in I. Bitran, S. Conn, O. Kokshagina, M. Torkkeli, and M. Tynnhammar (eds), *Innovation, the Name of the Game in Stockholm*, XXIX ISPIM, Lappeenranta University of Technology, LUT Scientific and Expertise Publications, pp. 1–11.
Remmen, A. (2001), 'Greening of Danish industry – changes in concepts and policies', *Technology Analysis & Strategic Management*, **13** (1), 53–69.
Stahel, W.R. (1982), 'The product life factor', in S.G. Orr (ed.), *An Inquiry Into the Nature of Sustainable Socities: The Role of the Private Sector*, The Woodlands, TX: Houston Advanced Research Center, pp. 72–96.
Weick, K.E. and R.E. Quinn (1999), 'Organizational change and development', *Annual Review of Psychology*, **50**, 361–86.
Weick, K.E., K.M. Sutcliffe and D. Obstfeld (2005), 'Organizing and the process of sensemaking', *Organization Science*, **16** (4), 409–21.
Weissbrod, I. and N.M.P. Bocken (2017), 'Developing sustainable business experimentation capability – a case study', *Journal of Cleaner Production*, **142** (4), 2663–76.
Zurn, J. and P. Mulligan (eds) (2014), *Learning with Lean : Unleashing the Potential for Sustainable Competitive Advantage*, Boca Raton, FL: CRC Press/Taylor & Francis Group.

PART IV

Staging interactions between research and innovation

9. Staging strategic enactment of front-end innovation[1]

Louise Brønnum and Christian Clausen

INTRODUCTION

Front-end innovation (FEI) is a term used to refer to activities aimed at developing new ideas and product concepts prior to product development activities, which in many cases are defined and processed through a stage–gate type model (Cooper 1990). During front-end activities, risk tolerance is supposed to be higher because the cost of failure is relatively low. Such innovative processes have been characterized as being explorative (March 1991), in contrast with formalized product development activities, which tend to be more exploitative. A recurrent discussion within innovation management concerns the possibilities and difficulties of combining explorative and exploitative activities within the same organizational structures (Tushman and O'Reilly III 1996) and the difficulties encountered by mature organizations that attempt to integrate innovative activities into more routine activities (Dougherty 2008; Jensen et al. 2018).

Traditional conceptual development activities based on formalized development processes with a strong focus on gate passing and checklists (Leifer et al. 2000) leave little room to explore different and more radical development opportunities and explorative mindsets and agendas. By defining FEI as an innovation concept focused on new concept development, scholars have attempted to establish a conceptual space aimed at supporting the explorative development paradigm (Koen 2004). Whereas Leifer et al. (2000) provided and described a variety of tools and approaches to FEI development, we suggest that the key issue is not methodological in nature. Rather, the focus for enabling FEI in a mature development company should be broadened to include how to navigate organizational development conditions such as stage–gate models, innovation strategies and development cultures.

In this chapter, we elaborate on how FEI is being staged as a development space to enable explorative and more radical innovation in a mature development organization with a highly structured and formalized development process. Analytically, we focus on the concept of a development space (Brønnum and Clausen 2013, 2015), which can be viewed as a temporary and discursive sociotechnical space (Clausen and Yoshinaka 2005, 2007; see also Chapter 2, this volume). We define a development space as a temporary space which specifically enables innovative product development in an organizational context. This definition sensitizes our attention to the ongoing political game (Dawson 2000) that must be played to gain support for particular development activities. Based on an in-depth case study of how development processes are organized and carried out in a mature company, we draw attention to how what we will term the *development constitution* is enacted (Weick 1988) in different ways, thereby staging and enabling different types of development spaces with different agendas. We explore how team members, particularly project managers, enact a diverse array of development conditions. The aim is to draw attention to and expand the repertoire of strategies for staging development activities available to team members and project managers engaged in FEI activities, and to extend understandings of how to stage more radical development spaces in mature organizations. We show how a deviating enactment strategy challenged dominant interpretations of development to enable more radical innovation. Our findings inform a framework for enacting strategic development by affording front-end concept developers in engineering design reflective navigational opportunities.

A STAGING APPROACH

We see the development space (Brønnum and Clausen 2013) as temporary and defined by the actors, political agendas, organizational elements, discourses, and actor-network collectives participating in and configuring the space (Clausen and Yoshinaka 2007). Viewing the development space in this way helps us understand development opportunities as outcomes of specific enactments (Weick 1988) of devices and actors situated both within and outside it. Development spaces should be seen as having rather open boundaries, as actors typically participate in more than one space at a time, and spaces transform as development activities unfold.

We use the term *development constitution* to refer to the common set of elements referenced by actors that influence how they perform FEI such as development strategy, organizational structure, the development model and development culture. This concept enhances our understanding of staging by highlighting key elements that must be navigated and enacted in order to enable development spaces, and how these are configured and attract organizational support in the form of management attention and resources. According to dictionary.com, the word *constitution* is defined as a system of basic principles according to which a nation, corporation or similar entity is governed, but is also used to refer to any established arrangement or custom. In the field of practised law, the word may refer to the composition of a set of entities, written or unwritten, which together govern a set of activities (Grey 1978). Likewise, scholars have used the term *company social constitution* (Kamp 2000; Clausen and Olsén 2000) to refer to the basic principles that inform the political navigation of interests and concerns in an organizational context. In turn such a social constitution is seen as the outcome of political processes and develops over time influenced by conflicts as well as their resolution and ensuing compromises. Similarly, we use the term development constitution to refer to a particular political practice or custom that links a number of entities specifically focused on (product) development within an organization and which inform political navigation. We are specifically interested in how such a development constitution informs the political navigation in the staging of a development space and how agency can be constructed through enactment of constitutional elements. As we will see in the following case, such elements are considered to include highly heterogeneous entities such as organizational devices like development models, key performance indicators (KPIs), innovation intents as well as the development organization, management positions, rules and practices, and the perceived development culture.

The development constitution is something which constitutes particular conditions of possibilities and can be seen as an obligatory passage point (Callon 1986) in the staging and enactment of innovation and development opportunities. In keeping with actor–network theory (ANT), these conditions or elements do not impact development on their own, but gain importance only through the ways they are enacted. Constitutional elements thus can be seen as nodes in a heterogeneous actor-network which connects to wider or even competing networks concerned with strategy, technology, markets and the like. A development constitution can perform in multiple ways as the diverse actor networks making up the constitution may be selectively enacted and mobilized.

Staging is about setting up and configuring a space, and considering which actors possess the necessary competences, mindsets and tools for the particular development activity (Andreasen et al. 2015). More broadly, the strategic orientation of staging aims to provide the condition of possibilities for network translation and innovation to happen (see also Chapter 2, this volume). In the context of FEI this includes the pursuit of development agendas, and the orchestration that enables development work to occur. Staging can be an intentional activity performed by key actors such as project managers or other management personnel, or as seemingly unintentional activity associated with particular devices such as stage–gate models, which may be used to enact particular agendas. Here, we investigate the particular configuration of a development space as a result of explicit or implicit staging moves, and how it depends on the elements selected to be enacted and the ongoing negotiations among participating actors. This process may vary in important ways, and we seek to understand whether and how the overall development space is configured for incremental or more deviating radical innovations.

The concept of *enactment* refers according to Weick (1988) to an actor actively participating in sensemaking by strategically enrolling and mobilizing other actors. Enactment involves strategically using agendas or identities to influence actors to engage in specific behaviours that result in a specific collective performance. In other words, 'Enacting involves shaping the world' (Weick 2009: 37). While Weick brings insight from organizational (social) theory, he refers to how enacted environments contain real objects, but that these objects are inconsequential until they are acted upon. From a material–semiotic ontology like ANT the notion of enactment similarly relates to the idea that actors build agency by relating to and translating the behaviours and actions of other actors and objects (Law 2002; Mol 2010). To enact objects is according to ANT to enact spatial conditions of possibilities, as objects may be enacted in a particular way (Law 2002) or may be framed differently (Mol 2010). We use the term re-enactment to indicate a re-considered and diverging enactment of similar conditions. As we will show, enactments and re-enactments characterize key strategic elements in the staging of development spaces and the implementation of different development agendas.

METHODS

We collected empirical data in 2013 as part of an in-depth case study of the business innovation (BI) department of a mature product development company in Denmark (hereafter, 'the Company'). The Company employed approximately 300 people responsible for developing dedicated analytical solutions for the agricultural and food industry. The overall aim of the study was to identify and understand the nature of FEI in a particular company context, and to outline FEI development opportunities. A qualitative research approach was appropriate, because fulfilling this aim required rich engagement with actors in situ.

Data were collected using ethnographic methods such as personal observations and interviews driven by open-ended questions to gain an understanding of implicit dynamics and latent knowledge (Boyle 1994). Over a six-month period, the first author spent two days per week at the development organization to become a familiar face in the company and to ensure confidentiality and trust when conducting interviews and observing daily activities. Data collection was an iterative process that combined formal interviews with observations of work processes (such as project meetings) and informal conversations at lunch. The field research focused on one ongoing project, Leap, that exemplified how the Company engaged in FEI. Leap was a version of FEI development focused on what the organization referred to as 'radical innovation'. During the study, the Leap initiative was formalized within the organizational structure, making it a particularly revealing case.

A total of 19 formal interviews were carried out with informants from different organizational divisions, including concept and product developers, business developers, one project manager, four section leaders, two vice presidents (VPs) and the executive manager (chief operating officer (COO)). The interviews prompted informants to reflect on what they viewed as important aspects of conditions for development (Spradley 1979). These interviews were later transcribed and analysed, along with archival data including company documents, development tools and processes, and corporate strategy statements.

During the analysis phase, we focused specifically on how interviewees engaged different organizational structures and the development model to gain an understanding of ongoing daily practices related to development and to identify potential opportunities specifically for FEI development. A workshop was held with informants to solicit feedback on initial findings and discuss opportunities and problems.

CASE STUDY: ENABLING FEI AT THE COMPANY

The Company was a Danish, tradition-bound, family-owned business with proud and classical engineering competencies. Known to be a bit conservative, the Company was a leader in a niche market for measurement instruments in which competitive dynamics were relatively well known. Over the last few years, a specific strategy for streamlining the product portfolio had informed the Company's development agendas. Revenue per unit sold had increased due to an emphasis on Lean project management principles focused on increasing efficiency in the development process with an orientation toward market launch, as well as addressing potential project risks before they became too costly.

The Company had a well-defined organizational structure for its development activities, which were handled by two divisions at the corporate headquarters. BI and product innovation (PI) were, from a management point of view, two equally important development functions which played two different roles. BI was responsible for bringing forth new ideas and concepts to feed into the product portfolio in order to maintain and expand market share, and PI was responsible for executing development from concept to launch. PI received initial concept ideas from BI in the form of concept reports, which it used to guide product development through the final stages. BI had approximately 30 employees, in contrast to the over 200 employees employed in the

PI division, indicating, in some respects, the relationship between explorative and exploitative innovation within the Company.

Terminology used by the Company reveals significant differences between the project modes in BI and PI. Even though the handover of concepts from BI to PI was a significant source of frustration, there seemed to be a clear mandate from executive management to maintain this practice in order to differentiate the development focus. The COO described the different orientations of the two divisions:

> When you are in a development function, then you will get a task to perform: here is a specification for this and this and this. It is like going to school. There you will get an assignment, but that is not like the concept developer. They will get a problem definition: the customer wants this and the ability to measure this. Okay, what do we do? Everything is up for grabs, so mentally you have to be able to work with very open-ended questions.

However, in practice, this difference is not as clearly defined. When a project is handed over to a new team, a lot more is at stake than a shift in development focus, including ownership of the idea, vision for the concept and trust in the potential for business and technology.

The Prevailing Development Model

The development model – performing as a front-end device (Clausen and Yoshinaka 2009; Bates and Clausen 2020) – played a key role in the staging of the development space, as it configured relations between actors and objects and defined progress in development projects. Company Way was a gate-based development model that served as the reference for the Company's development activities. It defined a number of obligatory gates (Cooper 1990) that needed to be passed in order to proceed with the project. The 'gates' in Company Way consisted of checklists for the proposed concept, as well as requirements for completing document templates used to hand over projects to the next phase. Company Way was constructed with a focus on the gates, and the terminology reflected this. 'Gate passing' was an important factor in performance evaluations of middle managers as well as assessments of overall division and section performance. BI 'owned' the first two gates (G1 and G2) whereas PI 'owned' the subsequent gates defined by Company Way. The head of BI was evaluated based on the number of specific projects that passed the G2 gate and were transferred to PI within a certain timeframe.

BI had two individual departments: business development (BD) and concept development (CD). Both departments focused on the early stages of development and innovation and handled FEI for the Company. As a department, CD was measured on the number of projects that passed gate G2 each year.

The development model therefore played an important role in understanding the context for early concept development processes. Because it was continuously referred to and enacted by project managers, developers as well as department and executive managers, Company Way was an important part of the development constitution. As just described, gate passing involved complying with gate checklists, but equally importantly, required convincing the review board as well as the chief executive officer (CEO) and COO (executive management) of a concept's potential. The review board was the formal committee governing the gates; although feedback provided by the CEO and COO was not part of the formal review process it was effectuated in the daily enactment of the development constitution. It functioned as a cultural and best practice, and many of the developers who had been in the Company for several years understood how to satisfy the review board by drawing upon arguments used to gain approval for previous projects. However, it was clear that executive management actors played an important role as they held a key mandate to review the potential of ideas. Likewise, executives could choose to give projects identified as too risky based on the gate checklist more time and slack to explore potential, thereby enhancing development within established structures, but with a larger margin for error. The COO described this function as his privilege and responsibility:

> I believe in informed absolute monarchy, understood in the way that I would be upset if I did not manage to listen to others; but it is not a club for discussion, and it is not a democracy. In the end I will make a decision and then we need to move forward.

This introduction to the primary way development was staged at the Company reveals the key role of the development model, gate-passing principles and strategies, best practices, and the identities and actual performance of executive managers as elements in the development constitution. The strong linkage of FEI activities to the product development model signified by the established rules for passing particular gates underscores that the dominant way of enacting the development constitution mainly staged an exploitative version of FEI. In the following section, we describe an emergent, more explorative version of FEI activities in the Company including the re-enactment of elements well beyond the prevailing development model.

Leap: Re-enacting the Development Constitution

Leap was a rather new FEI initiative focused on radical innovation. The actors engaged in the Leap project characterized themselves and each other as being driven by their need and desire to work with explorative development, which they claimed was not possible due to the way the development constitution had been enacted previously. The goal was to extend the Company's product portfolio instead of simply managing the current portfolio via what was internally labelled 'facelift' innovation. The Leap project is an interesting example of how the Company's development constitution was re-enacted by the project manager and his team, thereby setting the stage for a front-end development space with a configuration that deviated from common daily practice. Before we explore the re-enactment strategies adopted by the Leap project team, we believe it is useful to provide background information about Leap and the actor relations that made the project possible. Our method is to mainly follow the perspective of the project manager seeking room for more radical innovations, in the staging of an alternative development space.

The Leap manager emphasized that the earlier streamlining of the Company made a Leap-type initiative inevitable, as the strategic forecast revealed a need to extend the product portfolio and develop new markets and products. Another issue concerned the location and configuration of the deviating development space. Initiatives for pursuing radical innovation were not new in the organization. Earlier attempts to establish innovation hubs such as 'innovation greenhouses' or 'Company labs' all focused on R&D for new technical inventions to be fed into the development organization. These initiatives, however, proved to be too decoupled from the daily enacted development practice, which meant few, if any inventions actually became product development projects. Despite what one would expect, Leap was not anchored and initiated by BI, which was responsible for front-end development. Instead of being anchored in the formal front-end structure, Leap made progress as a side-line initiative enacting support from the more powerful downstream development organization. In the sections that follow, we trace how the Leap project created new development possibilities by continuing to comply with the current development constitution yet re-enacting key elements.

Re-enacting strategy

The Leap project explicitly deviated from the dominant practice of 'facelift' innovation. Its strategic aim was to extend the product portfolio by 'bringing new technology to new markets' through 'new-to-the-world innovation'. Although top managers created an opening for Leap by recognizing portfolio extension as a strategic priority, the Leap driver was passionate and eager to transform this vision into reality. While the strategy stated the need for a quantum leap it did not mention what the content of a project to meet such a need should be. First, he and his colleagues piqued management's interest by justifying the customer interest, potential revenue generated and technological potential of the team's idea. Specifically, they highlighted how the proposed technology could form the basis for a new technology platform for future products. They caught management's attention by speaking in terms of strategy, and by stressing the development of platform technologies. This was very much achieved by ways of framing, as expressed by a concept developer: 'It is not just a toxin, it is a sensor … it is not just soil analysis, it is mineral detection.' Moreover, they aligned their ideas with a particular aspect of the newly adopted Lean project management framework which focused on streamlining development activities.

Furthermore, proposing the new technology as a potential new core technology was viewed by the Leap project team as a strategically strong move, as it elevated the meaning of concept development to another level. A Leap concept developer explained:

> [This] is why it is now referred to as a technology that feeds into the core technology. So even though it does not fit into a fixed understanding, then we have accomplished something, we have raised the level. It may be put aside for some years, but then we can use it for something else.

Thus, members of the Leap project team were not limited by the scope of existing core technology modules (i.e., platforms), but had a mandate to explore new technology that could become one of the Company's core technologies in the future and align their work with the unit currently being responsible for core technologies, the Team Technology.

Early on, Leap's proponents developed a strategy tool referred to as 'innovation intents', which outlined opportunities and possible directions for new development. These intents were accepted by management, and were officially embedded into the Company's development strategy. 'We have identified some areas … suited for our company and line of business, … and now we have written them down in these innovation intents, and this is what we are working towards now' (Business developer). The innovation intents enabled members of the project team to justify the potential of a new technology based on official corporate strategy and goals, thereby making progress 'measurable' at gates. Such a proactive enactment of the Company's strategy seemed to be uniquely applied by the Leap project team.

Re-enacting the development organization

Having the support of executive management was critically important to ensure sufficient resources were allocated to radical innovations.

> Well, the decision always ends up being made by our top management, because if they do not grant the resources for the project then it is not going to happen, we can believe in the project, but at the end of the day, if there is no buy-in from their side, it is not happening. (Manager in PI)

The Vice President of PI responsible for the portfolio process had originally articulated the need for new products after Lean project management principles had been implemented. Having the Leap manager report directly to the Vice President and the COO provided a key opportunity to obtain management support and create a clear mandate. The Leap manager explained why the re-enactment of the management structure outside the normal chain of command was important for staging an environment that supports radical innovation:

> Leap would not have succeeded if I had not been able to go directly to the VP [of PI] who sits on the big box of money and has the mandate to make decisions. Had I had to go through some senior manager who would have some KPIs that did not revolve around this project … then the project had never been sold to the VP and I would not have gotten the funding and the possibilities I have.

The Leap manager further explained that this was an explicit navigational choice: 'There was a brief chatter about placing Leap in the core technology division, but I put my foot down and said that if that were to happen I would not be here.' Instead, it was decided that Leap should report to the Vice President of PI, the downstream division responsible for the exploitation of concepts delivered by the CD department in the BI division.

Another concern centred on the staging of cross-disciplinary collaboration. To address this concern, the Leap manager emphasized teamwork by establishing a mandatory morning meeting for project team members. Providing status updates, sharing new insights and identifying potential problems on a daily basis helped create momentum and a sense of teamwork. Frequent interaction encouraged diverse core technology experts to think holistically and enabled them to contribute to a common concept development framework. Invitations were even occasionally extended to experts outside the team who were viewed as potential 'recipients' of the technology. Inviting them to 'come and see what we are doing' helped lay the foundation for a future cross-disciplinary network around the concept proposal.

Re-enacting the development model: Company Way and KPIs

Many developers in the CD department felt restricted by the prevailing development model, Company Way.

> I am not the biggest fan of boxes and that it has to be in a certain way, and you have to use this template. I think it is killing the creativity, so I am certainly not a supporter [of Company Way] … I think it is developed for those who want to control how we perform. (Developer in CD)

Although Leap team members may have held similar views about Company Way, they enacted it in a different way to configure for different possibilities. The Leap manager explained:

> It is a necessity, as it gives us access to resources and visibility to management. To me a process is a communication tool, that is what it primarily is. When I say I have passed G1 then the CEO knows what I'm doing. Then I don't have to say anything else basically, and that is the visibility it provides.

A business developer further explained how the development model provided recognition and credibility, which was more important to Leap because risks were higher than in other front-end projects: 'We have ourselves chosen to follow that [development model] ... because we think ... getting transferred into PI with the same credibility as any other projects is important, and then we have told ourselves that we do not have to follow it all the way through.'

Also, the Leap team perceived and enacted the gate-passing process differently by emphasizing navigational possibilities which were not inscribed in the development model. Drivers of initiatives such as Leap explicitly considered the potential consequences of handing over a concept proposal with potential risks at G2. The VP of PI received the concepts at this stage, meaning he also evaluated the risks involved and therefore became an important potential ally. Gate passing thus was facilitated by pre-negotiation with management about the goals associated with each gate supported by devices such as innovation intents. This helped install a pull effect of expectations from potential developers who were waiting to receive projects after G2.

Re-enacting the development culture

The Leap project manager expressed an explicit intention to enact and establish a different mindset among the developers, because the mindset based on the dominant enactment of the development constitution left little space for radical thinking. Front-end development in CD was often characterized as focusing on eliminating risks from a potential concept. A concept developer characterized the development mindset in the prevailing development constitution as: 'if we can't kill it, it will survive', indicating that early development was focused on a concept's potential to fail instead of its possibilities. Similarly, the head of BD responsible for writing up the business case during the initial development phase described the main task of his division as 'killing the idea if it can be killed'.

In contrast, the Leap manager explicitly tried to build a new mindset for how development was considered, and how to stage a development space that enabled different types of ideas to develop:

> We have tried to develop a safe atmosphere ... we all know that if we just all say that there is a high risk, then it is not so dangerous to fail, because we all know that is an option, but we also know that this is where the potential is. (Leap manager)

Leap adopted the mantra that failing is acceptable, as long as you learn from it. The project manager argued that this competence building aspect of Leap created value for the core development organization, independent of the explicit result of the project itself. This is an example of how Leap enacted the development constitution in an alternative way by framing risks and failing as positive, as long as the organization learned from such experiences.

A FRAMEWORK FOR STAGING THE STRATEGIC ENACTMENT OF FEI

Figure 9.1 illustrates the Company's development constitution, the elements of which contribute to an understanding of how development is expected or ought to occur in the organization. This understanding changes over time due to new experiences, but retains best practices from previous projects. A development space for FEI appears as elements of the development constitution are enacted. As the case demonstrates, multiple enactments create multiple diverse development opportunities.

As illustrated in Figure 9.1, re-enacting the principles and elements embedded in the development constitution seems to enable the staging of alternative development space(s) for projects with different development agendas that deviate from established norms. We have shown how the Leap project staged radical innovation possibilities through the creation of deviating configurations of possibilities by:

- **Re-enacting strategy:** by articulating the teams' potential to contribute to portfolio extension and to core technologies the project manager enacted top management's more or less outspoken concerns as well as open formulations in strategy documents. Hereby the radical concept development became positioned as a strategic activity addressing and responding to long-term goals – and not just as generation of new ideas as part of FEI. By creating a new strategy device – the innovation intents – they established a mediator between strategic management and concept development. And by subsequently circulating and using these intents to actively negotiate project goals and making these measurable they created a pull effect for passing gates, replacing the normal barriers.

- **Re-enacting the development organization:** by rejecting the 'normal' references to the management of concept development, the project manager instead negotiated direct access to high level decision-making, positioning the project as a side-line activity to formal concept development in CD. By doing this he enabled access to and developed power by associating with the much more resourceful product development activities and created spaces for cross-disciplinary collaboration and interest from downstream actors waiting for exploitable concepts.
- **Re-enacting the development model:** by using the development model as a communication tool, thereby signalling compliance with existing dominant rules and continuing to emphasize the prevailing gate-passing process. In this way the project manager legitimated his radical concept development project by drawing on agency of the powerful model and thus keeping political contestation at a distance.
- **Re-enacting development culture:** by building a different mindset concerning the perception of development promoting a positive learning perspective on risk and failure as opportunities for organizational learning (Vinck 2017). By opposing and reframing the zero-failure culture the project manager and his team attempted to escape the prevailing development culture.

These re-enactment activities occurred as actors such as the Leap project manager navigated elements embedded in the development constitution, and reveal a repertoire of strategies for staging spaces for FEI to address a specific development agenda. More generally, our findings reveal the plural nature of enabling opportunities, and that multiple diverse enactments of a development constitution may co-exist in the same organizational context.

The staging of a development space with a specific configuration is achieved not as single staging moves but through the orchestrated enactment of diverse but interlinked material and social elements. These configuring elements include the circulation of devices such as innovation intents, combined with the navigation of the gate-passing process and networking strategies aimed at establishing alliances with top management and across expert domains enabling access to and support from downstream resources.

Our findings highlight the importance of intrapreneurial and navigational competences, as there seems to be a fine line between deviance from the development constitution and compliance with it. According to one of the developers, the Leap manager knew when to comply with company rules and when not to, so that he challenged embedded rules, practices and structures in just the right way. This became a matter of both complying with and (re)enacting

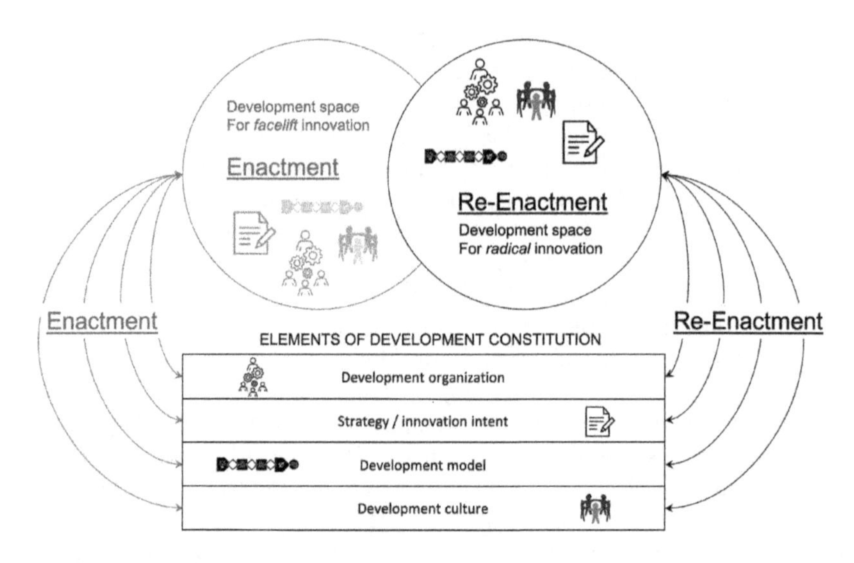

Note: This illustrates how elements of the Company's development constitution are enacted in order to stage a development space for facelift innovation or re-enacted to stage a development space for radical innovation.

Figure 9.1 Enactment of the development constitution

the constitutional understanding of development, navigating and translating these understandings into new meanings. Top management acknowledged these competences. The COO explained: 'There has always been room for someone like him [the current Leap driver] ... but you have to buy in yourself. I do not believe that you should create an organizational structure around it.' So, the agency involved in creating a deviating development space was forged around the Leap driver's political navigation of strategic openings created by management.

CONCLUSION

We have presented and analysed a case of how a manager responsible for FEI managed to re-enact what we term the development constitution of a company in such a way that his team seemed aligned with the core organization, but pursued a new development agenda. The implication is that in a mature development organization, there seems to be a navigational space for enacting rules, objects, conditions and understandings in different ways. By doing so, actors are able to stage new types of development spaces, thereby allowing for a variety of development projects with different innovation potential.

In this chapter, we have highlighted how the (re-)enactment of development rules, understandings, strategy objects and organizational structures and actors plays a key role in the staging of new development spaces for innovative thinking. As such it adds a political and organizational perspective to the existing repertoire of staging such as the inclusion and exclusion of actors and framing the meaning of the space. We have suggested the development constitution as a concept describing the conditions of development such as they are perceived by actors in the development organization. The most important elements of the development constitution identified in the described case are the perceptions of: the Company's development strategy, the development organization, the development model (Company Way), and the development culture as these elements are continuously referred to and subject to interpretation and framing. These elements are key to any political navigation of development opportunities and can be enacted in multiple ways. Re-enacting the principles and elements of the development constitution thus enable a development space to be staged in alternative ways and thus enable new opportunities.

The development constitution is a dynamic concept which reflects how the guiding principles and references for how to stage and enable FEI are constantly changing based on ongoing translations of meaning and enacted organizational resources and opportunities. The development constitution is a time-sensitive state of mind brought into existence as perceived by the actors involved in development practice. It represents a perspective on what influences front-end actors and creates agency. Because it reflects the past, shapes the present and has implications for the future, the development constitution may also be used as a reference for mindful deviation, such as in path creation (Garud and Karnøe 2001).

NOTE

1. The chapter is a highly revised version of a paper presented at the ICED15 conference (Brønnum and Clausen 2015).

REFERENCES

Andreasen, M.M., C.T. Hansen and P. Cash (2015), *Conceptual Design: Interpretations, Mindset and Models*, Heidelberg and New York: Springer International Publishing.

Bates, C.A. and C. Clausen (2020), 'Engineering readiness: How the TRL Figure of Merit coordinates technology development', *Engineering Studies*, **12** (1), 9–38.

Boyle, J.S. (1994), 'Styles of ethnography', in J.M. Morse (ed.), *Critical Issues in Qualitative Research Methods*, Thousand Oaks: Sage, pp. 159–85.

Brønnum, L. and C. Clausen (2013), 'Configuring the development space for conceptualization', *Proceedings of the 19th International Conference on Engineering Design (ICED13)*, Seoul: Design Society.

Brønnum, L. and C. Clausen (2015), 'Enabling front end of innovation in a mature development company', in C. Weber, S. Husung, G. Cascini, M. Cantamesa, D. Marjanovic and F. Montagna (eds), *Proceedings of the 20th International Conference on Engineering Design (ICED15): Innovation and Creativity*, Vol. 8, Milan: Design Society, pp. 235–44.

Callon, M. (1986), 'Some elements of a sociology of translation: Domestication of the scallops and the fishermen of St Brieuc Bay', in J. Law (ed.), *Power, Action and Belief: A New Sociology of Knowledge?*, London: Routledge, pp. 196–223.

Clausen, C. and P. Olsén (2000), 'Strategic management and the politics of production in the development of work: A case study in a Danish manufacturing plant', *Technology Analysis and Strategic Management*, **12** (1), 59–74.

Clausen, C. and Y. Yoshinaka (2005), 'Socio-technical spaces – guiding politics, staging design', *International Journal of Technology and Human Interaction*, **2** (3), 44–59.

Clausen, C. and Y. Yoshinaka (2007), 'Staging socio-technical spaces: Translating across boundaries in design', *Journal of Design Research*, **6** (1), 61–78.

Clausen, C. and Y. Yoshinaka (2009), 'The role of devices in staging front end innovation', in *Proceedings of the First International Conference on Integration of Design, Engineering and Management for Innovation IDEMI09*, Porto: University of Porto, pp. 1–8.

Cooper, R.G. (1990), 'Stage–gate systems: A new tool for managing new products', *Business Horizons*, **33** (3), 44–54.

Dawson, P. (2000), 'Technology, work restructuring and the orchestration of a rational narrative in the pursuit of "management objectives": The political process of plant-level change', *Technology Analysis and Strategic Management*, **12** (1), 39–58.

Dougherty, D. (2008), 'Managing the "unmanageables" of sustained product innovation', in S. Shane (ed.), *The Handbook of Technology and Innovation Management*, Chichester: Wiley, pp. 173–94.

Garud, R. and P. Karnøe (2001), 'Path creation as a process of mindful deviation', in R. Garud and P. Karnøe (eds), *Path Dependence and Creation*, Mahwah, NJ: Lawrence Erlbaum Associates, pp. 1–40.

Grey, T.C. (1978), 'Origins of the unwritten constitution: Fundamental law in American revolutionary thought', *Stanford Law Review*, **30** (5), 843–93.

Jensen, A.R.V., C. Clausen and L. Gish (2018), 'Three perspective on managing front end innovation: Process, knowledge, and translation', *International Journal of Innovation Management*, **22** (7), accessed 27 October 2018 at https://doi.org/10.1142/S1363919618500603.

Kamp, A. (2000), 'Breaking up old marriages: The political process of change and continuity at work', *Technology Analysis and Strategic Management*, **12** (1), 75–90.

Koen P.A. (2004), 'The fuzzy front end for incremental, platform and breakthrough products', in K.B. Kahn (ed.), *PDMA Handbook of New Product Development*, Hoboken, NJ: John Wiley and Sons, pp. 81–91.

Law, J. (2002), 'Objects and Spaces', *Theory, Culture and Society*, **19** (5/6), 91–105. DOI: 10.1177/026327602761899165.

Leifer, R., C.M. McDermott, G.C. O'Connor, L.S. Peters, M. Rice and R.W. Veryzer (2000), *How Mature Companies can Outsmart Upstarts*, Boston, MA: Harvard Business School Press.

March, J.G. (1991), 'Exploration and exploitation in organizational learning', *Organization Science*, **2** (1), 71–87.

Mol, A. (2010), 'Actor–network theory: Sensitive terms and enduring tensions', *Kölner Zeitschrift für Soziologie und Sozialpsychologie. Sonderheft*, **50**, 253–69.

Spradley, J.P. (1979), *The Ethnographic Interview*, Orlando, FL: Harcourt Brace Jovanovich College Publishers.

Tushman, M.L. and C.A. Reilly III (1996), 'Ambidextrous organizations: Managing evolutionary and revolutionary change', *California Management Review*, **38** (4), 8–30.

Vinck, D. (2017), 'Learning thanks to innovation failure', in B. Godin and D. Vinck (eds), *Critical Studies of Innovation. Alternative Approaches to the Pro-innovation Bias*, Cheltenham, UK and Northampton, MA, USA: Edward Elgar Publishing, pp. 221–39.

Weick, K.E. (1988), 'Enacted sensemaking in crisis situations', *Journal of Management Studies*, **25** (4), 305–17.

Weick, K.E. (2009), *Making Sense of the Organization, Vol. 2: The Impermanent Organization*, Chichester: Wiley.

10. Staging referential alignment in industrial–academic collaboration

Charles Anthony Bates and Joakim Juhl

INTRODUCTION

In this chapter, we investigate a collaboration led by the first author between tribological scientists and engineers from Luleå University of Technology (LTU) in Sweden and the Work Functions Division (WF) of the Danish owned Danfoss Power Solutions (DPS), a global leader in off-highway mobile hydraulics (Danfoss Group 2020: 7). The purpose is to describe and characterize micro-processes involved in the development of a new, environmentally friendly non-petroleum lubricant technology within a petroleum lubricant-based industry. Our staging focus is on how WF and LTU negotiated test specifications to document wear effects on machine components run with the new lubricant. The collaboration was initiated by LTU under the assumption that replacing standard hydraulic fluids with the new non-petroleum lubricant would require cooperation with the mobile hydraulic industry. Because DPS is a major technology leader and brand within this industry, the tests performed by the manufacturer were important for demonstrating the functionality and reliability of the new environmentally friendly lubricant.

Here, we explore how staging within the domain of technology development can be investigated and understood. Typically, scientific and industrial collaboration partners operate within different organizational and institutional settings and under different performance expectations (Juhl 2016). Although staging can involve many kinds of activities that produce new 'spaces' for opportunities within existing organizational frameworks, we direct our empirical focus specifically to a series of technology tests and how these can be understood in terms of staging. The tests were initially performed to demonstrate functional reliability for a new lubricant, but were later extended to generate deeper theoretical understandings of lubrication standards and attritional effects. Our framing of the term 'staging' therefore concerns the careful orchestration of technology tests and the collaborative environment necessary for its undertaking. In this space, we consider the test setup not only as a means through which staging was conducted, but also as an object of and the objective for the staging.

Our analytic intention is to conceptualize staging in relation to how the collaborative space is performed through and around the technology tests. Detailed descriptions of the work carried out before, during and after these tests, including the material agency of the tests themselves, are therefore instrumental to understanding the *what, how* and *why* of the staging activities. In short, it was the test setup that was staged through negotiations and interpretations around articulation of the technology tests. The 'why' is a bit more complicated, and requires situational analyses and questioning the performative nature of specific intentions behind the careful orchestration of the test setup. To conceptualize performances achieved through the technology tests, we adopt the perspective of 'innovation science', which denotes a 'domain of knowledge production … [wherein] academic scientists produce knowledge for commercial ends' (Juhl 2016: 136). Innovation science is inherently performative; 'in contrast to pure academic science that produces universal knowledge, innovation science produces knowledge that is particular to its intended application environment' (Juhl 2016: 137). Innovation science connects the relative universality of theoretical scientific knowledge with particular situated and technological conditions.

In this chapter, we examine how a distributed technology test setup was staged to produce coherent results that satisfied diverse organizational and institutional requirements in a knowledge exchange between the mobile hydraulic industry and the scientific field of tribology. To this end, we present a narrative describing how social, economic, organizational, technical and scientific elements were connected and made relevant to develop a test setup that could demonstrate the functionality and reliability of a new non-petroleum lubricant. An important finding is how standards were central to the collaboration's alignment of test parameters, interests and knowledge—particularly standards for petroleum lubricants, which shape the entire industry.

Technology Tests, Scientific Experiments and Standards

To investigate these dynamics, we draw on analytic resources from actor–network theory's (ANT's) semiotics and hybridity (Latour 1987, 1999), including Jasanoff's (2004) 'co-production' idiom, which is based on the understanding that normative and epistemic processes shape and constitute one another. To articulate the test setup's role as a coordinating object, we draw inspiration from Vinck's (2011: 25) idea of 'equipping intermediary objects': 'Through the process of equipping, new properties are conferred on the intermediary object and this contributes to the shaping of the design space and collective work.' Specifically, this chapter documents the process through which intermediary objects (mainly a test rig) were staged to progressively acquire properties of 'boundary objects' (Star and Griesemer 1989) and thus facilitate a many-to-many translation process of interessement. From this perspective, collaborating entrepreneurs were able to reduce local uncertainty while maintaining coordinated cooperation among allies through an 'indeterminate number of coherent sets of translations' (Star and Griesemer 1989: 390–1).

By treating the development of the test setup as a case of innovation science, we analyse how the collaborative partners staged what we term 'referential alignment' between the standards and objects of the mobile hydraulic industry and the representations and computations of the scientific discipline of tribology. Referential alignment is a term that we develop to characterize the quality of referential relationships that enables knowledge artefacts and processes to refer back and forth between different material, organizational, local and temporal settings. Referential alignment is fundamental to enable, and thus stage knowledge exchanges between distinct epistemic cultures (such as between industry and academia). Latour (1999) developed the 'circulating reference' concept to explain the amplification of a message that manifests itself throughout cascades of translations of matter–sign vehicles in scientific practice. Still, Latour's empirical demonstration of circulating reference in Boa Vista remained within the confines of established field sciences and their practices. Whereas the empirical site was new and unique, the methods underwriting the scientists' translation of the forest into a field laboratory fits the description of 'ready-made science' (Latour 1987). In ready-made science, norms for interpretation and evidential requirements are already established and considered stable enough for other matters to form translations and be built upon them. In contrast, when we examine staging of referential alignment in technology tests between scientists and industrialists, the performance resembles 'science-in-the-making' (Latour 1987), wherein facts and machines are still 'under-determined' and shared norms for interpretation and evidence have yet to be established.

Although the present case does not fall squarely within the realms of 'ready-made science' or 'science-in-the-making', the theoretical significance is that all collaborators simultaneously were trying to establish *what* they required the test setup to produce, and *how* the tests could be designed to deliver these outcomes as a shared and coherent epistemic foundation upon which the collaboration could operate. While determining what exactly needed to be tested, collaborators engaged in a complementary process to establish organizational and institutional expectations of the tests. In addition to the tests' techno-scientific content, the tests' normative context was also subject to ongoing staging in order to establish the tests' institutional and organizational affiliations, purposes and the performance criteria by which the test results would be evaluated.

RESEARCH METHOD AND BACKGROUND

Our theoretical framework builds upon ANT (Callon 1984; Latour 1987; Law 2009), which requires analyses to account for the agency of both human and non-human actors. This is especially important when considering spaces for technological development, where arrangements of test setups, industrial and scientific standards, and specialized measurement equipment all contribute to and condition the potential outcomes of processes of negotiation.

The first author led the collaboration in his role as Manager of Motor Engineering Technology (MMET) in WF. MMET played the role of 'auto-ethnographer', possessing 'prior knowledge of the people, their culture and language, as well as the ability to … "pass" as a native member' (Hayano 1979: 100) in the collaboration.

Shortly after the collaboration began, the MMET contacted the second author and extended an invitation to actively participate in the project, so that they might draw on innovation science to benefit the collaborators' reflective praxis. The empirical material draws on both authors' direct participation in translating results across the collaborating organizations, as well as academic papers and simulation models by WF, LTU or both, emails, test specifications, audio recordings of interactions, industrial brochures and standards, and ISO norms. All of these were discussed and (re)assessed with case participants prior to publication.

Methodologically, this chapter draws on 'laboratory studies' of ANT, which 'take into account the organizational variables linked to the allocation of resources, communication structures and relations between organizational entities' (Vinck 2010: 83). Viewing research as an enactment, insofar as 'realities are enacted with the discovery that they are enacted differently in different places' (Law 2009: 152), we draw on the notion of staging to describe and analyse how the constitution, configuration and transformation (Clausen and Yoshinaka 2009) of the actors and objects propagating the collaboration were (re)assembled in real time.

We present the notion of 'staging in situ' to sensitize towards translations that are enacted from within a commercialization collaboration by actors integral to the work by which both organizations subsist, as they (re)assemble objects in development processes while striving to retain and stabilize as much of their preceding work as possible. Although external consultants and/or researchers undeniably perform enactments within such collaborations, we consider staging in situ to be a skilled management practice different from that performed by researchers and consultants who are not embedded within the locally situated understanding of the objects they observe (and may or may not reciprocally manipulate).

Within such collaborations, objects underwrite the processes and goals they are situated to support (Vinck 2011). This is particularly true in scientific–industrial collaborations where principles of physics manifest through the results of experiments and tests. Although scientific experiments and technology tests both take place within controlled environments, experiments are performed to generate data to verify, challenge and build theory, whereas tests probe the functionality of technology and the usefulness of the models informing the particular design (Downer 2007). A significant consequence of this difference is that technology tests and scientific experiments are arranged to accommodate different targets. Tests typically refer to intended 'real-world' applications of technology, and are organized to alleviate issues that reduce the ability of results to demonstrate a specific functionality or reliability under field conditions. Experiments are designed to produce results that fit modelling parameters so that the underlying theoretical models can be assessed (Sismondo 1999; Winsberg 1999). Consequently, scientific and industrial data productions can reflect oppositional referential requirements in situations where continued collaboration requires referential alignment. Consequently, industrial technology tests and scientific laboratory experiments must be coordinated if they are to support a shared and coherent knowledge production process.

Within such collaborations, actors do more than just situate and manipulate objects towards data production:

> The equipping of intermediary objects changes the status and ontological properties of these objects and contributes to the shaping of the design space and work collective. By equipment, we mean any element added to intermediary objects enabling them to be connected to conventional supports and spaces of circulation. (Vinck 2011: 25)

The following narrative documents a process of *referential alignment* involving the transformation of an *intermediary object* into a *boundary object* which served as a centre stage for coordinating various agendas and interests through a progression of *equipping work*.

A DISTRIBUTED TEST SETUP

Since 2005, WF had worked closely with LTU to understand and simulate mechanisms influencing orbital motor performance and durability. This work included bespoke kinetic and mechanical simulations, ad hoc optical measurements, analyses of finishing processes, and multiple co-authored publications (Furustig et al. 2015a, 2015b, 2016). With over 100 million Swedish krona (SEK) in direct funding from industry, LTU was the university in Sweden with the strongest track record for industrial collaboration, and was internationally recognized as an authority in two crucial knowledge areas for WF: modelling and simulation of contact mechanics, and tribology of lubricated interfaces. Building on this history of collaboration, a Chaired Professor in the Faculty of Engineering and Natural Sciences approached the first author in his role as MMET to request industrial support in order to benchmark a new sustainable, renewable, nontoxic, water-miscible replacement for the *de facto* standard hydraulic lubricant.

Constituting the Test Setup

Together with associates from LTU, the Chaired Professor had recently established Sustainalube AB, a start-up company focused on developing, manufacturing and selling the lubricant. Sustainalube had already received Swedish public funding and prestigious innovation prizes, including Venture Cup Sweden and Swedebank Future Prize. Sustainalube had focused primarily on positioning the technology as a substitute for grease and chainsaw lubricants in the forestry industry, but was seeking to expand into new markets, particularly the mobile hydraulic market. This market places significant demands on hydraulic systems, with expectations for minimum maintenance and few breakdowns. Such demands pose two essential requirements: valid long-term stress-tests, and reliable data showing that the individual components comprising a hydraulic system can maintain high efficiencies over the entire component lifecycle, without adversely affecting the other components with which they interact.

The lubricant permeates all components in a hydraulic system and is the medium by which hydraulic energy is converted into mechanical energy. In mobile hydraulic systems, pumps, motors, proportional valves, steering units and transmissions typically draw fluid from a single reservoir. The same fluid is thereby commuted to propel or steer a vehicle, actuate cylinders or turn motors to generate work. Verifying the functionality and reliability of lubricants therefore requires a test setup that can circumscribe a wide range of applications and standards. Principal standards (for example, ISO 3448: 1992) classify lubricants according to viscosity grades, which indicate their compressibility and thermal expansion – two significant parameters for a hydraulic system's functionality and reliability. Other guidelines define lubricants according to their refinement or percentages of saturates and sulphur. Although these variants have significant effects on how equipment performs, the possibilities for interchangeability between these and non-petroleum lubricants are mostly unknown. This knowledge deficit has significant implications for guarantees between hydraulic machine manufacturers and their trade partners, thereby necessitating in situ testing and customized evaluations. To address these implications, Sustainalube needed to stage existing, albeit limited, industry-wide knowledge to validate its lubricant as a risk-free alternative to ISO 32 oil. LTU scholars were acknowledged experts in modelling theoretical effects of different tribological systems. Still, they required an industry representative to define and execute practical experiments to support these models—an arduous task for which WF was well-equipped.

Although DPS neither made nor sold lubricants, its products must accommodate conditions inherent to or engendered by them. Benchmarking effects of potential lubricants was therefore sensible. Moreover, 'sustainability' was a company-wide aspiration and provided additional justification for entering the collaboration. In accordance with principles of the UN Global Compact Initiative,[1] Danfoss Group (the primary concern under which DPS is included) had prepared a sustainability report every year since 2014 (Danfoss Group 2020). These reports highlighted how Danfoss conducted 'business responsibly and profitably, with a view to maximizing sustainable value creation for society' (Danfoss Group 2020: 2). Still, sustainable alternatives to ISO 32 remained largely non-existent. Additionally, environmentally friendly bio-lubricants were associated with large carbon footprints, high levels of toxicity and poor compatibility with the rubber and plastic components dominating the industry. The collaboration provided a unique branding opportunity for WF to demonstrate commitments to sustainability.

Configuring the Test Setup

To move forward, the Chaired Professor introduced the Sustainalube's Chief Technical Officer (CTO) to the MMET. In subsequent meetings, they discussed how the interests of their organizations could be accommodated. Interests included: affirming WF's position as a sustainability leader, verifying the functionality and reliability of Sustainalube's lubricant for mobile hydraulics, preparing articles for publication in academic journals and generating specialized knowledge across both organizations regarding the operation of lubricants in hydraulic systems. With preliminary interests established, the CTO was invited to WF to present his invention to the division's directors and senior managers. In preparation for this meeting, the MMET discussed the collaboration with the WF Motor Engineering Director. Drawing on correspondence with LTU, they formulated a statement for the meeting invitation that was rooted in notions of sustainability:

> To accommodate the hydraulic market's growing environmental and functional demands, researchers from LTU have designed a new prototype fluid with the benefits of water and the lubricating effects of oil. This lubricant is water miscible,

> nontoxic, made from renewable sources, has tuneable viscosity, and is incompressible, with good wear properties and excellent friction properties.

The WF Motor Engineering Director also allocated a Design Engineer, a Test Engineer, a Laboratory Technician and a 200-litre capacity test rig to the project for a ten-week period. Such test rigs were central to the company's technological development, where design or process changes underwent meticulous tests and evaluations in order to demonstrate their operational reliability and performance. This was also true for the WF–LTU collaboration, where the test rig was central to the production of empirical validations for the entire test setup. The test rig thus functioned as an 'obligatory passage point' (OPP; Callon 1984) from where and through which all other development activities transpired and were defined. Drawing on the dramaturgical metaphors of Chapters 2 and 15 in this volume, the test rig became the 'centre stage' upon which the collaboration would be orchestrated.

Ideally, the collaboration needed a test setup that would validate the functional reliability of the non-petroleum lubricant for the entire market. To this end, the effects of the different lubricants on the volumetric and mechanical efficiencies of an orbital motor had to be documented. Knowing the effects of different lubricants on the types and magnitudes of measured wear on sub-components was also necessary. To accommodate these actions, the MMET and Design Engineer settled on a method described in their recently published paper (Bates et al. 2020). They would mechanically measure the motors' key components using specialized equipment at WF before and after testing. They would also analyse data from scanning electron microscopes (SEM) at both organizations to evaluate types of wear inherent to the different lubricants. Finally, they would incorporate lessons from another project with LTU, utilizing specialized optical measuring equipment to create 3D surface roughness topographies of the gear sets before and after testing according to a new ISO standard that LTU had been utilizing since its inception, but which WF was just beginning to utilize. All motor components were categorized and labelled according to mechanical measurement results before sending them to LTU for 3D surface measurements. Components were then returned to WF for assembly and tests, remeasured (or SEM analysed) and sent back to LTU. This made it possible to compare similar sets of components across the petroleum and non-petroleum lubricants. The test design's careful preparation was intended to ensure sufficient referencing between the mechanical conditions of the two test runs, so that differences between the lubricants used in each test run would be the only 'relevant variable' to which differences in the recorded wear could be referred. Whereas the construction of references is a well-known practice within scientific work (Latour 1999), we observed this phenomenon in the context of technology development, where references were constructed to align material conditions between two situations and ensure juxtaposability between the recorded results of each. The knowledge produced was not universal in its applicability. Rather, the construction of references between material test conditions was situationally conditioned to the specific circumstances of the test setup (including the lubricants) through which operational performance was compared.

The test setup thus functioned as an object through which tasks, interests, and knowledge were coordinated among partners. Whereas Callon's (1984) idea of translation is based on the establishment of an OPP through which agendas and interests are realigned in order to support one main agenda, the test setup articulated, connected and coordinated plural agendas and interests through its referential infrastructure. As such, the test setup's coordinating role aligned more with Star and Griesemer's (1989) boundary object concept: articulating translation at multiple levels and leaving room to adapt to local needs while concurrently maintaining a common identity.

Having established the test setup's expected deliverables, WF still needed a test specification compatible with the allocated test rig and timeframe. This was easier said than done. The test rig had restricted flow and pressure capacities. Furthermore, conducting two test runs in an area where components could be significantly worn within the timeframe required testing motors at the limits of their intermittent specifications. This further reduced test possibilities to three displacement sizes (the volume of oil displaced by the gear set, in cubic centimetres) spanning two different motor types.

Once the motors were selected, the team needed to ensure that the two sets of motors were comparable, as described above. Comparability was central for creating referential alignment between the two planned test runs and ensuring that magnitudes and positions of wear could be correlated across the two lubricants.

Benchmark testing could now be initiated.

Transforming the Test Setup

The MMET, Design Engineer and Test Engineer were at LTU to discuss the results of their first test iterations and measurements with the Chaired Professor and CTO.

> **MMET:** The last time I was in this room was for cake, when Furustig got his doctorate.[2]

> **Chaired Professor:** Oh! Can WF still use his wear model? No one here has touched it since.

> **MMET:** Yes. Mostly the Design Engineer. The CTO asked him about it yesterday. We can use it to calculate the gear set's entrainment speeds at different flows, pressures and viscosities, so the CTO can molecularly tune the oil to solve our problem.

Their problem was complex. Two sets of motors had been measured and remeasured after tests with ISO 32 and Sustainalube lubricants. Unfortunately, Sustainalube gear sets showed extreme attrition, including adhesive, abrasive and multiparticle wear. Two gear sets had seized. Furthermore, the ten-week allocation for the test rig had expired. The CTO went to the whiteboard, speaking while writing equations:

> The oil film is too thin at EHL [elastohydrodynamic lubrication] contacts. Dowson and Higginson [1966] formulas look like this. The Design Engineer can give me entrainment speed and contact forces from the Furustig model. I also need ISO 32 properties. I can get the rest from ball-on-disc tests: friction, precise wear coefficients … Then I can tune the lubricant.

Ball-on-disc tests are an established scientific standard within tribology. Although test rigs are costly, and rare specialized knowledge is required to correlate results with in situ component tests, ball-on-disc results were the best approximation to theoretical values upon which practical simulations could be based.

> **MMET** [pacing]: Chaired Professor, we have 3D surface characteristics from the CTO and the specs for your discs. If we machine and supply representative discs and the CTO performs the tests and analyses the data, then I'm sure we can keep the test rig with Sustainalube oil, at least for the rest of the year. Their results contribute to future modelling capabilities … But how does LTU get paid?
>
> **CTO** [interjecting]: Sustainalube's funding application included ball-on-disc. We have the means to pay LTU.
>
> **Test Engineer:** If WF needs the friction and wear coefficients, the lab should support this. Not just for the LTU know-how—that's huge—but also our focus on sustainability.
>
> **MMET:** Correct. And don't underestimate the academic articles this will generate. These support our position as industry leader. We should be driving the industry.
>
> **CTO:** What if I tune the oil to a different viscosity, where it can outperform ISO 32?

Albeit ambitious, the request was problematic. Orbital motors were a single component in a system comprising multiple components connected to the same tank. The industry standard was 35 cSt viscosity, roughly equivalent to ISO 32 at 40 degrees Celsius.

> **MMET:** No. Sorry. Mobile hydraulic is a conservative industry geared to 35 cSt. It's the only way in.
>
> **Test Engineer:** Look I'm new. But why 35? If it's better at 14 or 40 or whatever, customers will want it? Our motors are run hard. Steering units last forever. Motors are where the wear is?
>
> **Chaired Professor:** Yes, but ISO 32 is the culmination of, what, 100 years of ad hoc development? It wasn't chosen; it evolved. It won't take 100 years for something to replace it, but it's the criterion … Sustainalube will pay for ball-on-disc. So yes, the CTO can tune the lubricant. But is there a test area within 35 cSt that could help move the bar?
>
> **MMET:** Re-dimensioning our tests and lubricant to a different application area? … Maybe. But we'll need a new benchmark.

Although a comparison between the two test runs produced empirical data on differences in operational properties between the two lubricants, the data were not directly commensurable with theoretical parameters such as those used in the Furustig model. Dowson and Higgison formulas are based on parameters that are not directly measurable in situ. To establish these theoretical parameters for the lubricant, a dedicated laboratory ball-on-disc setup was required to control and remove as many variables as possible and ensure that output data was commensurable with the theoretical parameters the CTO needed to re-tune the new lubricant. WF tests had established an *industrial context* in which the new lubricant was made comparable with the standard ISO 32 oil. Adding LTU ball-on-disc equipment to the mix intended to engender a *scientific context* in which the lubricants and the WF-fabricated discs could function as industrial objects, whose friction properties could become theoretically knowable. By creating a coherent infrastructure of references, the facilitation of these industrial and scientific tests would establish referential alignment among industry standards, ISO 32 oil, the new lubricant, and the tribological models serving as a relevant explanatory resource. It was precisely this referential alignment (across industrial settings, academic fields, theoretical models, new and old lubricants and standards) that established conditions for the exchange of knowledge and know-how between industry and academia. In turn, this referential alignment was conditional for both the theoretical tuning of the new lubricant *and* its industrial reliability.

Back in Denmark, the MMET met with the Engineering Director and the Laboratory Manager. His message was clear. Although promising, the non-petroleum lubricant was not fully developed: simulations needed to be made, ball-on-disc tests needed to be completed. The test specification also needed to be changed to accommodate an ideal application. Sustainability was not low-hanging fruit.

The Laboratory Manager was supportive. The collaboration advanced condition monitoring competencies which supported future lab investments. Although the Engineering Director agreed that co-authored papers between WF and LTU would support WF's status as a sustainability leader, he was chiefly concerned with the capacity of ball-on-disc tests to improve WF simulation capabilities. Weighing the costs of testing against new theoretical and practical knowledge and branding potential, the BU Engineering Director, Laboratory Manager and MMET agreed that Sustainalube benchmark tests should continue. The panel would be made available for an additional 16 weeks.

DISCUSSION: STAGING REFERENTIAL ALIGNMENT

The WF contribution was to facilitate production of new knowledge regarding the lubricant's performance within a controlled industrial test environment that emulated 'real-world' conditions. The WF test lab granted the material conditions for establishing referential alignment between industrial objects, operational practices and ISO standards. In return, LTU provided the new lubricant technology and state-of-the-art scientific testing equipment, with data output conforming to established scientific standards within tribology. This constellation of hybrid academic and industrial data production with mutual reference to the same test setup was essential to stage a reference-borne knowledge infrastructure through which members of the hydraulic industry and tribology experts were able to exchange knowledge, know-how and success criteria. Collaboratory success in this case can be seen as depending on the ability to facilitate referential alignment and thereby (re)situate knowledges, methodologies and material objects across diverse organizational and institutional requirements.

Staging sustainability was a central objective for the collaboration and initially served as the qualifying success criterion towards which the fulfilment of different visions and resources were aligned. Although supporting its reputation as a sustainability leader had originally piqued WF's interest in the collaboration and helped constitute the network, sustainability benefits were insufficient to sustain the network in the face of unexpected test results. Fortunately for the collaboration, the initial objective of demonstrating WF's commitments to sustainability could be re-staged as supporting new opportunities presented by ball-on-disc tests. Whereas ball-on-disc tests were deemed necessary *objects* for Sustainalube to tune its lubricant, they became an *objective* for WF, providing them with entirely new opportunities to understand gear set mechanics and tribological interfaces and to expand modelling capabilities.

Still, Sustainalube's and LTU's objectives remained unchanged: Sustainalube needed to provide a functional and reliable alternative for ISO 32 oil, and LTU wished to maintain and improve its position as a world leader in industrial research collaboration.

Staging the Sustainalube lubricant proved more complex, as it could only be connected and made relevant for (and by) the test setup through a one-to-many OPP—namely, functional reliability as per ISO lubrication standards. Unfortunately for Sustainalube AB, these standards proved to be specific in their application to petroleum-based lubricants for which and by which they had been derived. The molecularly situated nature of the standards severely limited the conditions of possibility by which the functionality and reliability of the new lubricant could be validated. This resonates with Hardstone et al. (2006: 70–1), who proposed that:

> Where the scale and scope of activity become too large for direct scrutiny, simple forms of communitarian trust may come in to play based upon presumptions of reciprocity, broadly shared norms and established repertoires of behaviour and rooted in experience of repeated performance.

Just as oil is the medium that permeates a hydraulic system, the ISO 32 oil standard permeates the entire hydraulic industry, where its repeated performance has produced normative expectations that idealize its characteristics as the dominant technology in a market 'too large for direct scrutiny'. As pointed out by the Chaired Professor, these characteristics were neither designed nor chosen, but rather incrementally shaped and refined over the last century through use within the simultaneously emerging industries to which it served as a common reference point. The OPP by which the new lubricant could be (re)assembled into the test setup had the effect of supporting ISO 32 dominance.

To complicate matters further, the initial function of the WF test rig was that of an intermediary object still oscillating between translating goals into results and instigating action and negotiation (Vinck 2011). It was the constituting and configuring work that 'equipped' the test setup into possessing more substantial boundary object characteristics as it gradually became more integral to the operations of the entire collaboration.

The addition of ball-on-disc to the test setup would make it possible to identify and quantify the parameters by which the new lubricant could be tuned and made relevant, providing new means to accommodate demands defined by the ISO lubrication standards. Perhaps more importantly, ball-on-disc tests could provide theoretical values by which WF could expand and improve its practical simulations, as well as provide LTU with fundamental understandings of tribological interactions which could be connected to practical experiments. As such, ball-on-disc was not just assembled into the test setup in pursuit of centralized goals. Rather, it became a goal unto itself. Whereas other collaboration targets shaped preliminary and final ideas about what the test setup should perform—namely, augmenting WF's status as a sustainability leader, supporting LTU's status as a top industrial collaboration partner, and providing commercialization opportunities for Sustainalube—incorporating ball-on-disc into the test setup was both a *means* to accomplishing these targets and a separate *end* for WF, providing access to theoretical coefficients for future simulations.

CONCLUSION

Our primary focus in this chapter has been to provide an empirically substantiated theorization of the 'what, how and why' of staging in a case involving technology tests and the associated industrial–academic hybrid production of knowledge. We have invoked the staging metaphor to sensitize readers to the ongoing work that practitioners put into constituting, configuring and transforming diverse material and epistemic conditions in order to create and maintain a referential infrastructure across different situated settings. An important part of this type of staging is the negotiation and coordination of normative expectations and performance criteria in order to define a shared space for the different collaborators.

As a key object, the test setup was critical in the early connection of partners around 'common performance criteria'; paradoxically, the results of the test setup required the network (in the form of partners, resources and the setup itself) to be staged around a new problem definition and new objectives. Here, the possible means of tuning a new non-petroleum lubricant were dependent upon industry-wide standards geared towards maintaining the superiority of petroleum-based lubricants. Consequently, the collaboration had to be reconstituted, reconfigured and retransformed around a new theoretical problem the scientists excelled at solving, rather than the initial and more practical demonstration problem for which the industry partner had been approached for assistance. Within this process, maintaining referential alignment across the different organizational settings meant that knowledge built by the collaborators had to be broken down and replaced with questions whose answers required a new agenda. As the principal subject of the staging considered here, the test setup became more than a means for demonstrating the functionality and reliability of the new environmentally friendly lubricant. It also became a dialogic tool for negotiating established notions of validity.

Finally, this chapter documents the upstream constitution and configuration of an intermediary object to perform as a boundary object (that is, a referential infrastructure) as its comparability and commensurability were shaped across diverse stakeholders. Here, an intermediary object (the test rig) was initially staged as an OPP to support translation and alignment. Through this process of constituting the preliminary establishment of involved interests and configuring, through new knowledge production and its translation into reference construction and alignment of material conditions, the test setup represents a way to transform intermediary objects into boundary objects, and to ensure juxtaposability between the results of each, in supplement to the process documented by Vinck (2011). Conceptualizing this as a process of referential alignment could be the subject of future work.

NOTES

1. 'A voluntary initiative based on CEO commitments to implement universal sustainability principles and to take steps to support UN goals' (https://www.unglobalcompact.org/about, retrieved 8 May 2020).
2. Dialogues were translated from Swedish or Danish, edited for clarity and approved by those cited.

REFERENCES

Bates, C., H. Broe-Richter, C. Bendlin and P. Ennemark (2020), 'Case study: The effect of an amorphous hydrogenated carbon-coated gear-wheel on a hydraulic orbital motor's efficiency over time', *Proceedings of the Institution of Mechanical Engineers, Part J: Journal of Engineering Tribology*, **234** (3), 320–33.

Callon, M. (1984), 'Some elements of a sociology of translation: Domestication of the scallops and the fishermen of St Brieuc Bay', *The Sociological Review*, **32** (1_suppl), 196–233.

Clausen, C. and Y. Yoshinaka (2009), 'The role of devices in staging front end innovation', in *Proceedings of the First International Conference on Integration of Design, Engineering and Management for Innovation IDEMI09*, Porto: FEUP Faculty of Engineering and Management of the University of Porto, pp. 1–8.

Danfoss Group (2020), *The Sustainability Report 2019*, Nordborg, Denmark: Danfoss Group.

Downer, J. (2007), 'When the chick hits the fan: Representativeness and reproducibility in technological tests', *Social Studies of Science*, **37** (1), 7–26.

Dowson D. and G. Higginson (1966), *Elasto-hydrodynamic Lubrication: The Fundamentals of Roller and Gear Lubrication*, Oxford: Pergamon Press.

Furustig, J., A. Almqvist, C.A. Bates, P. Ennemark and R. Larsson (2015a), 'A two scale mixed lubrication wearing-in model, applied to hydraulic motors', *Tribology International*, **90**, 248–56.

Furustig, J., R. Larsson, A. Almqvist, C.A. Bates and P. Ennemark (2015b), 'A wear model for EHL contacts in gerotor type hydraulic motors', *Proceedings of the Institution of Mechanical Engineers, Part C: Journal of Mechanical Engineering Science*, **229** (2), 254–64.

Furustig, J., A. Almqvist, L. Pelcastre, C.A. Bates, P. Ennemark and R. Larsson (2016), 'A strategy for wear analysis using numerical and experimental tools, applied to orbital type hydraulic motors', *Proceedings of the Institution of Mechanical Engineers, Part C: Journal of Mechanical Engineering Science*, **230** (12), 2086–97.

Hardstone, G., L. d'Adderio and R. Williams (2006), 'Standardization, trust and dependability', in K. Clarke, G. Hardstone, M. Rouncefield and I. Sommerville (eds), *Trust in Technology: A Socio-technical Perspective*, Dordrecht: Springer, pp. 69–103.

Hayano, D.M. (1979), 'Auto-ethnography: Paradigms, problems, and prospects', *Human Organization*, **38** (1), 99–104.

Jasanoff, S. (ed.) (2004), *States of Knowledge: The Co-production of Science and the Social Order*, Abingdon-on-Thames: Routledge.

Juhl, J. (2016), 'Innovation science: Between models and machines', *Engineering Studies*, **8** (2), 116–39.

Latour, B. (1987), *Science in Action: How to Follow Scientists and Engineers through Society*, Cambridge, MA: Harvard University Press.

Latour, B. (1999), 'Circulating reference: Sampling the soil in the Amazon forest', in B. Latour, *Pandora's Hope: Essays on the Reality of Science Studies*. Cambridge, MA: Harvard University Press, pp. 24–79.

Law, J. (2009), 'Actor network theory and material semiotics', in B.S. Turner (ed.), *The New Blackwell Companion to Social Theory*, Hoboken, NJ: Wiley-Blackwell Publishing, pp. 141–58.

Sismondo, S. (1999), 'Models, simulations, and their objects', *Science in Context*, **12** (2), 247–60.

Star, S.L. and J.R. Griesemer (1989), 'Institutional ecology translations and boundary objects: Amateurs and professionals in Berkeley's Museum of Vertebrate Zoology, 1907–39', *Social Studies of Science*, **19** (3), 387–420.

Vinck, D. (2010), *The Sociology of Scientific Work*, Cheltenham, UK and Northampton, MA, USA: Edward Elgar Publishing.

Vinck, D. (2011), 'Taking intermediary objects and equipping work into account in the study of engineering practices', *Engineering Studies*, **3** (1), 25–44.

Winsberg, E. (1999), 'Sanctioning models: The epistemology of simulation', *Science in Context*, **12** (2), 275–92.

11. Staging with objects: translation from technology to product development

Charles Anthony Bates

INTRODUCTION

The successful transition of inventions from technology development to product development is a well-known challenge in both industrial practice and management scholarship. Mainstream management literature provides numerous frameworks and process models targeting industrial practitioners who seek to effectively span transitions from invention to commercialization. Nevertheless, this body of literature often neglects how technology managers and engineers practically negotiate the opportunities and limitations which arise *through* interaction with these frameworks and process models.

In this chapter, I propose that problematizations concerning the suitability of a technology for product-specific applications are (re)negotiated across heterogeneous networks as the developing technology encounters new actors and changing notions of functionality and reliability. Furthermore, I submit that objects play significant roles in how such negotiations are staged, and that sensitization towards this process can benefit technology managers and engineers tasked with such translations. These are not radical propositions. Building on the 'laboratory studies' of Latour and Woolgar (1979), Vinck (2009: 1) argues that by focusing on technical reality, 'a different vision of technology will emerge—a vision that technicians should find easy to understand because it will be based on their day-to-day life'.

Following an industrial case, I consider the transition from a *technology development project* (TDP) to a *product development project* (PDP) as a process of alignment across distinct (albeit mutually shaping) networks tasked with separate goals. Henceforth, I define technology development as a collection of diverse activities whereby the functionality and reliability of inventions are negotiated and improved across specialized networks to enable them to be incorporated into predetermined product-specific applications. As such, a TDP differs significantly from the structured PDP into which it feeds. PDPs focus on commercialization and are based on the assumption that technologies with sufficiently high levels of functionality and reliability can coalesce with a viable business case through structured customer interactions (see process models described by Cooper 2008).

This chapter responds to a limited understanding of the roles of objects within engineering management practice and how they are staged by managers and engineers. In mainstream literature, the transition from technology to product development is viewed as an exercise of idea identification, selection and maturation in a mostly orderly and mechanistic process (see for example, Cooper 2008; Florén and Frishammar 2012; Markham et al. 2010; Verworn et al. 2008). This scholarship claims to serve engineering work, yet does not always embrace or address the contents of such work. In mainstream management models, it is typically a management team coordinating how work is 'staged' (defined in Chapter 2 of this volume as the inclusion/exclusion of actors, material and symbolic objects and concerns in a space and the construction of boundaries defining the space of development). Often, this staging is framed as occurring through the proxy of a project manager or 'champion' equipped with an assumedly stable set of commercial criteria and a team of supporting specialists. Still, management literature mostly ignores the fact that 'models are not neutral but offer certain framings, contribute translations and act as sensemaking devices' (Clausen and Yoshinaka 2009: 1), and neglects how more mundane objects like prototypes mutually shape commercial criteria.

In contrast, this chapter focuses on the negotiations whereby technology managers and engineers accommodate changing scopes of action, with a focus on how a variety of objects help and impede translations as they are 'configured, stabilized and facilitated' (Clausen and Yoshinaka 2009: 4) throughout the staging process. Objects are interesting because they enable us to look at staging within engineering practice, as a complement to the management-centric staging found in management models. I use the term 'staging with objects' to highlight the reciprocal nature of what diverse actors do with shared objects, and how such exchanges (un)intentionally (dis)align the networks into which the objects are assembled.

This chapter is organized as follows. In the next section, I briefly introduce my analysis of 'staging with objects', including my research method and background information about the company. Then, I present a case highlighting the roles of objects involved in a project's translation from technology to product development. I follow this with a discussion of how objects are staged across heterogeneous networks, and conclude by elaborating on staging with objects as an engineering practice.

ANALYSING STAGING WITH OBJECTS

I reflect on 'staging' as a practice by which *entities*, defined as networks composed of humans and non-humans, as well as material and immaterial objects, are materialized and (re)assembled to perform in processes of translation. An engineering report is such an example, where finite relationships between standards, prototypes, measurements, test rigs and analytic methods are assembled and materialized in a written document. My reflections are rooted in classic actor–network theory (ANT) and the work of Clausen and Yoshinaka (2009: 2), who considered 'the role which devices play in the managing of FEI [front-end innovation], with inspiration from science and technology studies ... to examine and discuss devices that intervene at the front end'. Clausen and Yoshinaka considered staging from an academic perspective by observing, participating and interviewing to make sense of industrial practice. I exploit their notion to reflect on my own practice, combining insights as a technology manager employed in industry with an academic analytic perspective.

In this chapter, I view innovation as a process of translation: a movement between the practicalities of invention and commercialization, where the networks comprising such practicalities are inevitably displaced. To substantiate these aspects of translation, I align with scholars in science and technology studies (STS). Within this community, the notion of 'heterogeneity' describes how the 'stability and form of artifacts should be seen as a function of the interaction of heterogeneous elements as these are shaped and assimilated into a network' (Law 1987 [2012]: 113). Heterogeneity thereby emphasizes 'material practices that generate the social' (Law 2009: 148), to describe '*how* relations assemble or don't' (Law 2009: 141). Furthermore, I consider technology and product development to be two distinct, albeit mutually shaping endeavours with shared social and material elements and dynamic ends. To analyse such translations, Vinck et al. (1996) propose using intermediary objects to help identify 'actors and characterise the forms of organisation and coordination, and the agreements binding them' (Vinck 2012: 91). Vinck et al. (1996) differentiate two types of intermediary objects: 'commissioning objects' translating goals into results, and 'mediating objects' instigating action and negotiation. This represents a network perception of the nature of objects, where 'objects are an effect of stable arrays or networks of relations' (Law 2002: 91). An important implication of this view is that 'innovation' takes place across and through numerous networks, where objects are designed and assembled to perform in interaction with other networks.

Research Method and Background

As an 'auto-ethnographer' (Hayano 1979: 100), I possessed 'qualities of often permanent self-identification with a group and full internal membership' in the networks considered. Building on these relations, I gathered empirical data from (inter)company communications, standards and specifications, journals, and audio recordings. In my role as a TDP manager, I also co-authored the patents, test specifications and whitepaper described herein. Analytically, I draw on ANT, which: 'treat[s] everything in the social and natural worlds as a continuously generated effect of the webs of relations within which they are located. It assumes that nothing has reality or form outside the enactment of those relations' (Law 2009: 141). To avoid confusion, I use 'object' to denote an actor-network that is stable or 'punctualized'[1] and 'elements' to denote the (un)stable 'arrays or networks of relations' (Law 2002) comprising these objects, such that each object is considered to be a heterogeneous network of elements.

The setting for this study, Danfoss Power Solutions ApS (DPS), is a global leader in the industry in which it operates, providing complete hydraulic systems for the agriculture, infrastructure and material handling markets. In the Work Functions (WF) division, a TDP spans diverse, specialized networks, such as the division's engineering, production, sales, purchasing and leadership teams, and its customers. Here, inventions are meant to feed into future applications, the specificities of which are not always known and where yields contributing to these specificities are identified and defined within the confines of the TDP. Even though assumptions regarding future uses of technology take root in a finite and heterogeneous network of objects and relationships which are assembled and verified according to company and industrial norms, these assumptions are still challenged as the technology moves towards the well-defined application-specific considerations of a product development process such as that outlined by Cooper (2008). Consequently, new actors and contextualizations, such as market players with diverging product programmes and architectures, can alter existing notions of functionality and reliability, thereby destabilizing the objects and relations upon which initial assumptions rest. In the next section, I present a specific case documenting the transition of a TDP to a PDP.

AN INDUSTRIAL CASE OF TRANSLATION

In this case, I managed a TDP with the aim of maturing a novel technology that would make hydrostatic steering safe and comfortable at increased speeds. At one point, most aspects of the technology were evaluated as having high readiness for commercialization in multiple *technology readiness assessments*

(TRAs; see Mankins 2009). Hoping to capitalize on this progress, the business unit vice president met with the engineering director and the TDP team to align the product strategy and discuss 'how technology development activities could more quickly transition into product development'.

At this juncture, the 'launch goals and strategy' were still unclear. Although the sales team had facilitated several technology demonstrations at company and customer test tracks, the 'scope and sequence' of possible vehicle systems into which customers might implement the technology remained uncertain. Ascertaining this scope and sequence requires specific commitments between the division and its customers, and these commitments presuppose high levels of technology readiness before solutions can be discussed. Furthermore, implementing novel technologies into complex off-highway mobile equipment often requires alignment with *other* significant changes to vehicle architectures in the customer pipeline. Although the new technology was supposedly far enough along to seek the customer commitments and alignments necessary to confirm a preliminary business case, the TDP and sales teams had not requested top management's approval to initiate such negotiations with customers. Referring to the TDP team's 'impressive TRA results', the vice president was ready to begin these negotiations, and asked the sales team and me to develop a whitepaper describing preliminary 'pilot customer considerations' for presentation to the division's top management.

Over the following weeks, I set up and facilitated recurring 'whitepaper meetings' with the TDP team, the engineering director, the sales director and sales managers to document pilot customer considerations in the contexts of: (a) the preliminary business case from which the TDP had sprung; (b) general characteristics of the hydraulic steering market, including their relevance to the technology; and (c) any necessary customer-specific considerations, commitments and alignments we could identify. After each meeting, I (re)composed sections of the whitepaper and sent them to the sales director and sales managers for review and revision. The original text, revisions and comments were then (re)negotiated in plenary discussions, after which I would (re)write sections of the whitepaper and (re)send them for review or revision together with any supplementary notes, documents or pictures.

Aligning Readiness with Commercialization

Early in the project, the TDP and sales teams agreed the technology project would initially span two functional principles encompassing numerous variants. The first principle was purely hydraulic. The second principle was an electro-hydraulic solution. Accordingly, the TDP team developed and submitted nine patent applications covering the variety of vehicle architectures deemed relevant via experiments, customer discussions and demonstrations.

Although both principles had been discussed and tested at length, both internally and externally with customers, the TDP team was mostly focused on a single variant of the first principle, with intentions to develop other principles and variants later.

Prior to meeting with top management, the sales manager revealed unexpected news. While reviewing specific considerations, commitments and alignments for the whitepaper, a pilot customer wished to consider implementing the new technology across-the-board. Consequently, the functionality and reliability of the technology would need to span the customer's entire relevant portfolio before commitments to application-specific implementations could be made. Furthermore, a portion of previously unconsidered vehicles required a third type of steering unit (hereafter, the third principle). This meant three principles needed to be developed instead of two, and an implementation plan had to be created for all three before applications of the technology could be identified and considered. Although the third principle had been drafted earlier (via patents, experiments and discussions with customers), technical complexity and sales forecasts had pushed it to the project's fringes. It existed only as a proof-of-concept. The following exchange[2] highlights the challenges faced by the team.

> **TDP Mechanical Hardware Specialist:** We didn't plan to mature two, let alone three principles simultaneously. Maybe there's synergy between the second and third principles, assuming our focus can span their consecutive development. But that's a serious undertaking! It doesn't coincide with the speedy transition from technology to product development.
>
> **Sales Manager** [interjecting]: But it's the transition required for any realistic business case. Our customer wants plans to develop and deliver all three before committing.
>
> **TDP Manager:** This places extra, unexpected demands on our technology readiness. It negates our plan to develop the different principles piecewise. This does not coincide with what we promised top management.

Indeed, the division's top management found the news surprising and disappointing: 'Why wasn't the second principle further along?' They also expressed concerns about the third principle elaborated in the whitepaper. According to project assumptions, the technology was also feasible for this tractor market segment: 'So where was the plan for variants distinguishing this market?' In closing, the division president defined a new agenda:

> You've made impressive progress with aspects of the technology, but relationships between patents, design and test activities, and commercialization are still unclear ... A necessary market remains undefined ... The timing of the new variant is a surprise. Maybe we needed to achieve a certain readiness before mapping the

> market … but from now on, commercialization must drive development. The TDP manager will continue to lead development until we identify a PDP manager … In the meantime, sales will focus on developing a robust business case.

A few weeks later the new PDP manager joined the project, working closely with the sales director and his managers to redevelop the business case while I facilitated tests and analyses in preparation for the next readiness assessment. After a four-week overlap, the PDP manager took over coordination of development activities, with a mandate to transition into a PDP when the business case was complete.

Re-aligning Self-alignment

A month later, I met with the PDP manager to discuss his progress. He said: 'As you know, we thought the new technology was most suitable for a specific hydraulic system. We didn't think its other functionalities were easily separated from self-alignment. Placing much of the market temporarily out-of-reach.' I did remember. These were key considerations in patents for the second and third principles. Moreover, not all of the principles were cost-effective across the different hydraulic systems. The PDP manager continued:

> Not all customers want the first principle at high-speed. Not always because of cost, for some OEMs [original equipment manufacturers] it's also their niche … The good news is, we can probably deliver third principle functionality by reducing self-alignment characteristics of the first principle.

This was big news. A variant for the third principle market was not developed when I left the TDP. What happened? In tractor tests of the newest prototype, specialists noticed poorer than expected performance of self-alignment with other functions mostly performing as they should. The mechanical hardware specialist explained:

> It was buried in the prototype's valving system. A manufacturing error changed a dimension by a few microns. Revisiting the patented hydraulic diagrams, making calculations and analysing test data, we realized it was possible to tune the first and second principles' self-alignment, to where they perform as a *de facto* third principle steering unit with negligible effects on other functions … Simulations are promising … We've ordered prototypes and tests.

Later, the PDP manager told me:

> If the specialists are right, a business case is close. We still need to develop the third principle for implementation with pilot customers, but we can tune self-alignment in the design and validation process towards developing the solution … Our readiness

is back on track! I'm working with sales and top management on launch considerations for the new business case.

DISCUSSION: STAGING WITH OBJECTS ACROSS HETEROGENEOUS NETWORKS

In this section, I draw on the narrative to consider how objects are staged in translations from technology to product development. I focus on the performance of objects in enabling and coordinating translations, with an emphasis on how the deliberate circulation of objects by specific collective actors such as the top management, sales and TDP teams (un)intentionally (dis)aligned the networks into which the objects were assembled. To that end, I consider the roles played by four objects in four different *movements*, or intentional courses of action towards defined objectives.

Figure 11.1 illustrates four objects—the TRA, whitepaper, tuned self-alignment[3] and business case—that were chronologically staged in the four different movements. Each object is visualized as an object box containing elements comprising the object's network. These networks are not comprehensive. Drawing on methodological rules set by Latour (1987), I include elements from a 'network in action', considering only the observed performance of an element within a specific movement. When an object box includes a previous object among its network elements, I also include elements from the previous object box which were reopened in negotiations during the movement. For example, the whitepaper object box includes the TRA object as well as the preliminary business case, patents, working principles and steering characteristics which were part of its network and were explicitly renegotiated during the movement. If an element is not carried forward to the next object box, it implies that the element was (re)punctualized in the previous network.

Figure 11.1 shows network elements, together with the actor-networks involved in shaping them. Note that the actor-networks which shaped the elements changed across movements. Although the top management, TDP and sales teams were all involved in negotiations concerning the preliminary business case, patents, working principles and steering characteristics in the whitepaper object box, these elements were negotiated exclusively by the TDP team in the tuned self-alignment object box. Dashed arrows in the object boxes show the order in which elements were (de)stabilized in negotiations between movements. The four movements are shown as curved arrows above the object boxes. Note that each movement begins with an existing object and ends with a new object. The actor-network initiating a movement is shown at the start (left) of each arrow. The actor-network responsible for a movement's negotiations is shown at the centre of each arrow; for example, the TDP and sales teams coordinated movement between the TRA and whitepaper. The

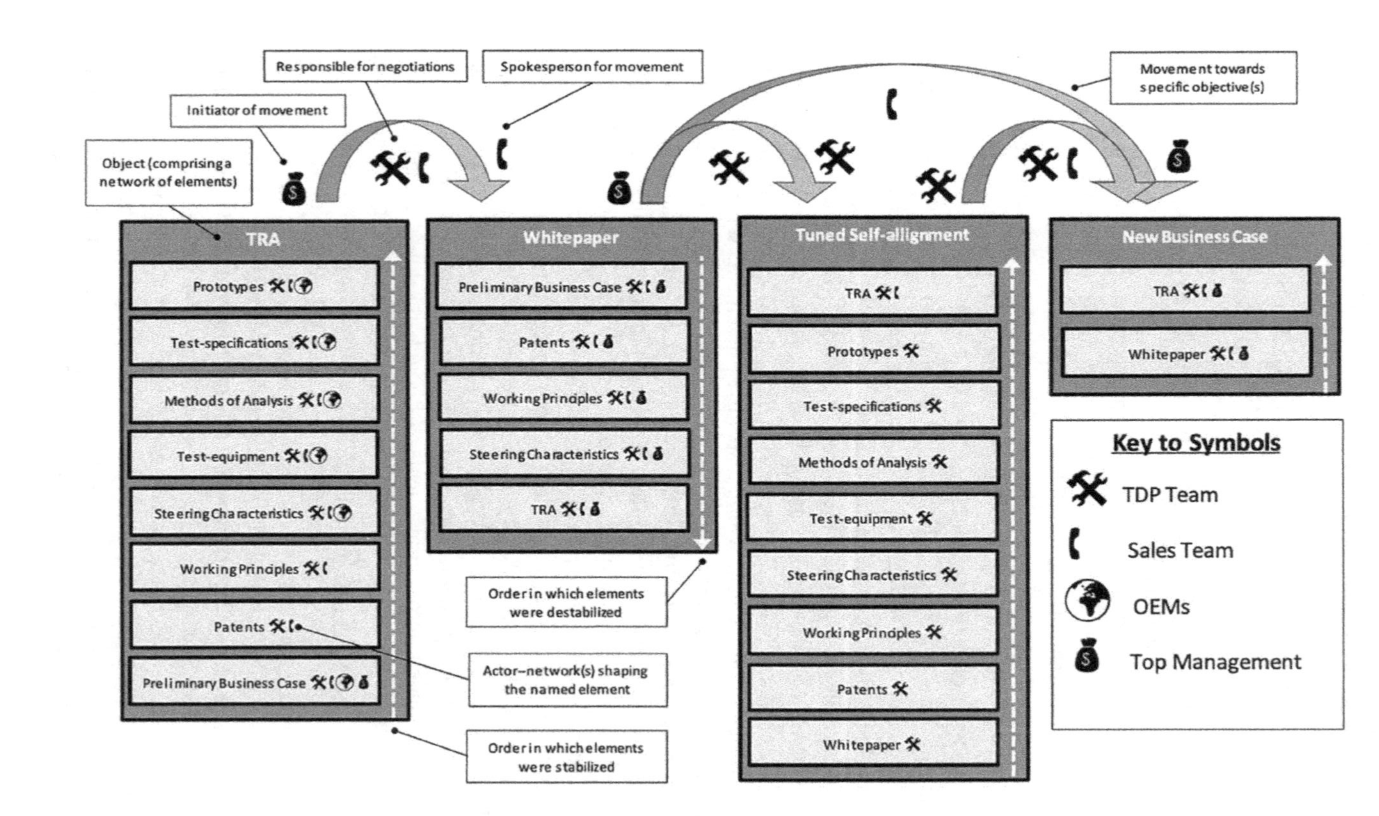

Figure 11.1 Four movements in the translation from TDP to PDP

actor-network with a company mandate to define outcomes of a movement within the firm (the spokesperson) is shown at the end (right) of each arrow. Consider the first movement from TRA to whitepaper: the actor-network known as the sales team spoke for the whitepaper—a mandate rooted in the sales team representing the 'voice' of OEMs within the confines of the firm. As such, the TDP, sales and top management teams are both actor-networks and spokespersons.

The First Movement: From the TRA to the Whitepaper

The TRA was intended, and appeared, as a stable object for enabling the technology to move towards product-specific applications. Even so, the TRA was destabilized when encountering new demands engendered by the increased (albeit necessary) involvement of customers and managers in the transition from TDP to PDP. The initiative through which the preliminary business case was intended to be updated and validated (the whitepaper) faltered; paradoxically, the preliminary business case (a product of earlier customer interactions) was made obsolete by the very object commissioned to support it.

The TRA generated a 'finalized number' (Callon and Muniesa 2005; see also Bates and Clausen 2020) which could represent the foundational apparatus of the technology across contexts, without needing to bring the entire apparatus with it. This enabled a new attempt to commit customers to an implementation strategy, the whitepaper. Still, the foundational apparatus was itself undermined by translations of new customer demands that the attempt fostered. As customer requirements were redefined through interactions with the whitepaper, the functionality and reliability of the technology lost credibility with the top management team and the preliminary business case became obsolete. The first movement thus resonates with earlier STS observations on how seemingly mature inventions are destabilized in interaction with end-users (see Fleck 1988).

The Second Movement: From the Whitepaper to the New Business Case

The second movement was instigated by the top management team in response to the destabilizing effects of the first movement on the TRA network. It can be summarized as the TDP and sales teams' engagement with top management to accommodate the new notions of functionality and reliability fostered by deeper engagement with OEMs. When presented with the whitepaper, the top management team issued a new mandate: 'from now on, commercialization drives development'. This wording is interesting, as it illustrates different perspectives on translation, seen from the vantage points of PDP and TDP teams. In product development, network building ideally begins with specific (assumedly stable) customer perspectives. In technology development, network building begins with the functionality and reliability of inventions for specific (assumedly relevant) applications. The new business case had to accommodate both.

The top management team articulated a necessary and overarching objective: commercialization (per the whitepaper) would drive the technology towards the new business case. Still, a direct connection between the whitepaper and new business case was absent. The *means* by which the network elements of the TRA object could be re-stabilized (or superseded) to accommodate findings in the whitepaper were neither known nor defined. An *interim movement* between the whitepaper and new business case was necessary. The TDP's reconceptualization of 'self-alignment' would serve as the object through which the whitepaper and a new business case could be connected.

The Third Movement: From the Whitepaper to Tuned Self-alignment

Considering the mandate that commercialization would drive development from that point forward, one may assume that activities preceding the whitepaper lacked a commercial focus. Such an assumption oversimplifies the development process. Activities did not lack a commercial focus; rather, the commercialization criteria that drove technology development were dynamic (see Fleck 1988).

As a milestone for moving industrial technologies through development phases, technology readiness was only relevant to *specific* end objectives (in this case, commercialization criteria). Although pilot customers actively supported the TDP and sales teams in defining the development process, unforeseen demands to implement the technology into new applications resituated the scope and sequence of a possible launch. The TRA network was destabilized because the whitepaper specified new commercialization criteria that differed from those guiding the technology's development.

To accommodate unexpected market expectations the TDP team needed to reconsider the working principles and steering characteristics driving the project. The discovery and recontextualization of a *functional flaw* ('poor self-alignment') provided a means to re-open these punctualized networks, address the destabilizing aspects of new market demands, and finally stage 'tuned self-alignment' as a *functional feature.* This was possible because self-alignment was conceptualized in such a way that new validations could be reconfigured from existing elements; that is, the specifications, test setups, calculations and simulations supporting self-alignment were quasi-stable. Already designed and implemented by the TDP team to validate a variety of problematizations, network elements could be reassembled to validate new functional principles. Staging self-alignment in a new context (as tuneable) made concerns regarding the functionality and reliability of the technology tangible and malleable. This enabled the TDP and sales teams to negotiate new problematizations with familiar (albeit re-dimensioned and resituated) network elements. In other words, 'a certain kind of reality was [again] recognizable within the conditions of possibility' engendered by the whitepaper (Law 2009: 149).

The Fourth Movement: From Tuned Self-alignment to the New Business Case

This movement is associated with restabilizing the TRA and punctualizing tuned self-alignment. Although incomplete when this chapter was written, the fourth movement yields interesting insights into the process by which objects are staged towards punctualization. Consider Figure 11.1, where the TRA and the whitepaper are situated in the new business case object box as stable network elements. This reflects how the TDP and sales teams were successfully staging the TRA network to accommodate revised navigational considerations borne by the whitepaper. But notice that self-alignment does not appear as a separate entity in the new business case object box, even though it remained an important network element of the TRA object. Successfully staged by the TDP team as a response to dynamic commercialization criteria and leveraged to re-stabilize readiness, tuned self-alignment had been punctualized. Its interim role in the translation was apparently fulfilled.

Tuned self-alignment, which had been performed as part of a deliberate strategy 'to create a durable network' could now be 'translated whole and "black boxed" into the [new] web' where how the object works 'is of little direct interest' (Law 2009: 148). Although staging tuned self-alignment *facilitated* a necessary movement towards the new business case, the object will likely continue to lose visibility (and direct relevance) as the translation nears completion. Successfully staged, tuned self-alignment was resituated as one of many imperceptible critical technology elements comprising the network of steering characteristics – itself a punctualized element of the TRA network.

STAGING WITH OBJECTS AS AN ENGINEERING PRACTICE

Defined by the highly institutional process model employed by the company, the constitution of the transition space was largely a given. Working from assumptions permeating a preliminary business case, the TDP team was expected to establish a smooth transition from technology to product development. They were, from the outset, keen to avoid 'throwing the concept over the wall' between TDP and PDP.

This study illustrates a transition from invention to commercialization encompassing messy processes of interwoven translation, including a diverse array of shifting actors and objects that the orderly mechanistic processes depicted in mainstream management literature do not adequately capture. Focusing on the role of objects in staging these types of movements, I have conceptualized *staging with objects* to make sense of seemingly chaotic processes, while emphasizing the reciprocal nature of what actors do with objects as they are circulated. As such, the notion of staging with objects challenges the idea of a 'great designer' manipulating diverse elements to known ends (Law 2002). In the case presented here, actors rarely, if ever, managed to successfully stage objects single-handedly. Rather, translations were characterized as responding to possible means of action by identifying and connecting with objects already in circulation, always in cooperation with other actors and dependent on technical acumen. Movements from invention to commercialization resembled less a scripted play than improvisational theatre.

The configuration of the transition space was based on a translation of concerns voiced by actors (the TDP, PDP and sales managers, the vice president and president) who served as spokespeople, representing the perspectives of their networks to translate strategic challenges and opportunities in different ways. Although I was the primary actor responsible for staging the transition through the first movements (in my role as the TDP manager), managing the movement from TDP to PDP became a larger political concern as agency to initiate, negotiate and communicate the results of movements shifted between the TDP, sales and top management teams in unexpected ways (see Figure 11.1). This illustrates how the staging required to move a concept from TDP to PDP does not reside in a single actor, but becomes a matter of negotiation among shifting (f)actors. One implication of this is the difficulty in identifying a single actor or network capable of managing the entire transition process. In the final movement, my role in making sense of the perspectives of the top management team, sales team and customers, and altering technological objects was taken over by the PDP manager, who also responded to what were often destabilizing actions in his attempts to stabilize an actor-network constituting a viable product concept. Tracing these efforts, staging implies numerous and different actions, including assembling and coordinating intersecting networks, constructing and/or assembling knowledge objects, and (re)circulating and (re)defining these objects through a dialogic process. The process described in the narrative exemplifies a successful, albeit meandering, orchestration of object creation/reframing and network translation.

Such an orchestration can hardly be seen as the simple outcome of an explicit and goal-directed staging process. Rather, it is an outcome of open, interactive processes spanning diverse actors and objects. A key finding is how shifts in staging efforts across organizational perspectives and competences were effectuated by the definition, development and circulation of intermediary objects oscillating between commissioning and mediating roles. Accordingly, competing concerns and strategies for how to stabilize or destabilize these objects becomes a key aspect in how movements from technology to product concepts and network translations are staged. As such, this translation from technological innovation to product concept cannot be adequately understood as the transition of a singular 'product concept' through or between presupposed situations. The objects considered here *all* performed as intermediary objects involved in the construction, circulation, (de)stabilization and eventual punctualization of an invention as it moved towards commercialization.

Surprisingly, the principal object involved in restabilizing the TRA network in response to the whitepaper was borne by a functional *flaw* staged as a functional *feature*. Here, the deliberate redefinition of tuned self-alignment (through cross-disciplinary collaboration and reframed observations) served as a key translational move to reconfigure and eventually stabilize a product concept. This move re-established the TDP team as a key actor in the overall translation process and points at the importance of not punctualizing (or black boxing) technology concepts too early.

Moreover, attending to the capabilities of a cross-disciplinary project team to reassemble technological concepts during *ad hoc* movements between invention and commercialization seems of key importance. Whereas definitions of technology elements were rooted in assumptions from a preliminary (and later destabilized) business case, their functional conceptualizations were flexible enough to survive dis- and re-association. I hypothesize that the obscure, albeit permeating, role of the tuned self-alignment object, deep in the 'machine room' of the already punctualized network in which it was integrated, played a significant role in the staging efforts to turn an unsuccessful (destabilizing) translation into a successful (stabilizing) translation. The lesson seems to be that translation of the broader TDP network depended on deep technological knowledge accumulated in the TDP network, coupled with abilities to identify and reassemble an actor-network behind punctualized objects and to develop new conceptual objects for (re)circulation in dynamic contexts.

In conclusion, I present the following points to help technology managers and engineers utilize staging with objects as an engineering practice:

- The agency to situate and define objects is a complex process of negotiation and does not reside in a single actor. Objects are reciprocally (de) stabilized by the effects of actors' changing positions as they navigate changing situations.
- Staging with objects requires increased attention to how objects perform across intentional (commissioning) and unpredictable (mediating) roles.

This necessitates conscious reflection over how objects are accepted, contested and changed in interaction with different perspectives.

- Attention to processes of staging with objects can help technology managers and engineers articulate and accommodate fluid demands from top management teams who set overarching objectives, exercise control over resources, and provide mandates for action, but are otherwise not directly involved in solving technical issues in translations from invention to commercialization.
- Viewing 'commercialization criteria' as products of ongoing negotiations between diverse actor-networks can provide more effective strategies (*means of action*) than viewing these criteria as stable, punctualized entities that guide how a technology is developed.

NOTES

1. Punctualization: 'The process by which complex actor-networks are black boxed and linked with other networks to create larger actor-networks' (Callon 1991: 153).
2. Quotes were translated by the author from Danish (where necessary), edited for clarity and brevity and discussed with cited individuals.
3. As an object, tuned self-alignment was the materialization of conceptualizations through which a specific functionality was achieved. A heterogeneous network including mathematical simulations supporting its reliability, and all specifications for fabricating, controlling and assembling components into a prototype.

REFERENCES

Bates, C.A. and C. Clausen (2020), 'Engineering readiness: How the TRL figure of merit coordinates technology development', *Engineering Studies*. DOI: 10.1080/19378629.2020.1728282.

Callon, M. (1991), 'Techno-economic networks and irreversibility', in J. Law (ed.), *A Sociology of Monsters: Essays on Power, Technology, and Domination*, Abingdon-on-Thames: Routledge, pp. 277–303.

Callon, M., and Muniesa, F. (2005), 'Peripheral vision: Economic markets as calculative collective devices', *Organization Studies*, **26** (8), 1229–50.

Clausen, C. and Y. Yoshinaka (2009), 'The role of devices in staging front end innovation', in *Proceedings of the First International Conference on Integration of Design, Engineering and Management for Innovation IDEMI09*, Porto: FEUP Faculty of Engineering and Management of the University of Porto, pp. 1–8.

Cooper, R.G. (2008), 'Perspective: The Stage-Gate® idea-to-launch process—Update, what's new, and NexGen systems', *Journal of Product Innovation Management*, **25** (3), 213–32.

Fleck, J. (1988), *Innofusion or Diffusation? The Nature of Technological Development in Robotics*, PICT Working Papers No. 4, Edinburgh: University of Edinburgh, Research Centre for Social Sciences.

Florén, H. and J. Frishammar (2012), 'From preliminary ideas to corroborated product definitions: Managing the front end of new product development', *California Management Review*, **54** (4), 20–43.
Hayano, D.M. (1979), 'Auto-ethnography: Paradigms, problems, and prospects', *Human Organization*, **38** (1), 99–104.
Latour, B. (1987), *Science in Action: How to Follow Scientists and Engineers through Society*, Cambridge, MA: Harvard University Press.
Latour, B. and S. Woolgar (1979), *Laboratory Life: The Social Construction of Scientific Facts*, Beverly Hills, CA: Sage.
Law, J. (1987), 'Technology and heterogeneous engineering: The case of Portuguese expansion', reprinted in W.E. Bijker, T.P. Hughes and T. Pinch (eds) (2012), *The Social Construction of Technological Systems: New Directions in the Sociology and History of Technology*, Anniversary Edition, Cambridge, MA: The MIT Press, pp. 105–28.
Law, J. (2002), 'Objects and spaces', *Theory, Culture & Society*, **19** (5/6), 91–105.
Law, J. (2009), 'Actor network theory and material semiotics', in B.S. Turner (ed.), *The New Blackwell Companion to Social Theory*, Hoboken, NJ: Wiley-Blackwell Publishing, pp. 141–58.
Mankins, J.C. (2009), 'Technology readiness assessments: A retrospective', *Acta Astronautica*, **65** (9–10), 1216–23.
Markham, S.K., S.J. Ward, L. Aiman-Smith and A.I. Kingon (2010), 'The Valley of Death as context for role theory in product innovation', *Journal of Product Innovation Management*, **27** (3), 402–17.
Verworn, B., C. Herstatt and A. Nagahira (2008), 'The fuzzy front end of Japanese new product development projects: Impact on success and differences between incremental and radical projects', *R&D Management*, **38** (1), 1–19.
Vinck, D. (2009), *Everyday Engineering: An Ethnography of Design and Innovation*, Cambridge, MA: MIT Press.
Vinck, D. (2012), 'Accessing material culture by following intermediary objects', in L. Naidoo (ed.), *An Ethnography of Global Landscapes and Corridors*, London: IntechOpen, pp. 89–108.
Vinck, D., A. Jeantet and P. Laureillard (1996), 'Objects and other intermediaries in the sociotechnical process of product design: an exploratory approach', *The Role of Design in the Shaping of Technology*, **5**, 297–320.

PART V

Staging experimentation and learning

12. Staging interventions in resource-intensive practices and related energy consumption levels

Charlotte Louise Jensen

INTRODUCTION

The wellbeing of humans and other species now and in future generations is vulnerable to the effects of climate change, and radical changes are urgent and required (IPCC 2014, 2018). Despite significant efforts by the EU as well as national and municipal governments to reduce domestic energy consumption over the last 20 years, traditional problem framings, which typically rely on a mix of rational consumer choice models, efficiency measures and information-based behavioural change theories, have failed to deliver anticipated reductions (EEA 2013; EC 2019). Given the ongoing ecological and environmental crisis and continued pattern of severe resource depletion, we need to explore other ways of understanding and facilitating change in production and consumption systems (see, for example, the 2015 United Nations Climate Change Conference, COP 21, and the Paris Agreement). The radical change processes necessary to address the climate crisis are unlikely to come about if we continue to build on traditional theoretical concepts of change that are informed by rational consumer choice models and certain types of efficiency measures which currently dominate policy strategies related to changing consumption patterns (see further evidence of this in Foulds et al. 2017; Labanca and Bertoldi 2018; Shove 2017; Sovacool 2014).

Because conventional approaches offer inadequate potential for radical change, new problem framings—that is, new ways of (theoretically) defining problems and how to address them—are needed to conceptualize, understand and engage with the challenges of high levels of energy consumption. Building on the idea that theories or problem framings produce particular forms of knowledge and, in turn, ideas about the spaces in which change can or will occur (Jensen et al. 2019), problem framings serve as a starting point for providing different conceptualizations with the potential to facilitate change in resource-intensive consumption patterns.

Adopting a practice theoretical approach, the European Network for Research, Good Practice and Innovation for Sustainable Energy (ENERGISE) has developed and tested new ways of framing the problem and challenging current resource-intensive practices and related energy consumption levels. A practice theoretical perspective on change offers different explanations and thus different spaces for change. Whereas behavioural change theories often frame individuals' attitudes, behaviours and choices as central drivers of consumption patterns, and therefore as targets for change (Shove 2010), social practice theories ask altogether different questions about the nature of consumption and why certain patterns exist (Shove and Walker 2014). For social practice theorists, collective practices take centre stage as they seek to understand why norms, ideas, traditions and habits are stabilized or challenged by different sociomaterial configurations of everyday life. Following this line of argumentation, ENERGISE thus understands interventions in norms, routines and habits related to domestic consumption areas to be a matter of intervening in relationships between particular configurations of images, skills and material arrangements in specific moments of time and space, what Shove et al. (2012) termed *practice configurations*.

With this in mind, ENERGISE has developed a Living Lab approach whereby practices are both the unit of analysis and focus of intervention. This approach can be described as a set of steps in a staging process: particular actors are invited onto the stage at different times, and different spaces are configured to facilitate particular types of deliberations and interventions. In this chapter, I present and discuss: (a) the different steps in ENERGISE Living Labs and how they are staged; (b) the types of results and outputs being produced through staged interventions inspired by practice theory, compared to interventions that rely mainly on rational consumer choice models; and (c) what a practice theoretical understanding contributes to the concept of staging.

ENERGISE LIVING LABS AS A PRACTICE-ORIENTED AND EXPERIMENTAL CHANGE PROCESS OF STAGED SPACES

ENERGISE and its scope has been described in detail elsewhere (see for example, Jensen et al. 2018). One of ENERGISE's core focus areas was resource-intensive practices related to heating and laundry. In more than 300 households across eight European countries, the team staged interventions not only to reduce energy consumption, but also to understand and change existing practices related to those domains. Two ENERGISE Living Labs (ELLs) were established in each country with approximately 20 households each, and all ELLs followed similar procedures and design. Before involving the participating households, the ENERGISE team put extensive time and effort into designing the overall framing of the ELLs as well as selecting the areas of household consumption to be targeted. For a detailed description of the ELL design and design process, please see Laakso et al. (2019).[1] Here, I present and discuss the process implemented in the Danish ELLs.

The households were recruited in collaboration with representatives of Roskilde Municipality, who helped distribute the recruitment survey in two areas of the city. The sites were sampled to ensure that residents with a mix of socio-demographic profiles would be reached. The recruitment survey also sampled respondents in such a way that it would be easy to identify types of houses, socio-demographic composition of households, levels of energy consumption and access to heating systems and laundry machines.

All recruited households were visited by the ENERGISE researchers, who introduced the project, the timeline and the different steps in the ELL process. As part of the first meeting in September 2018, all households were asked to start (and keep) a diary of how often they did laundry, what type of laundry they did, and why. In addition, they were asked to document the temperatures in their living and sleeping areas and to indicate whether the temperatures were comfortable or not. The households submitted this baseline information once a week through an online survey. The baseline period lasted four weeks for laundry and an additional three weeks for heating. In October 2018, the ELL researchers visited the households again. This time, the process was slightly different between the two ELLs: in ELL1, the households were visited individually, and in ELL2 members of the households gathered in a larger, focus group-like meeting. For the purpose of this chapter, I will not elaborate on the reasoning for this difference, other than noting its existence. In both cases, the households were asked to describe their heating and laundry practices, particularly in terms of social expectations, material contexts, traditions and childhood experiences.

At the end of the second meetings, the ELL researchers revealed the challenges for the next seven weeks: a four-week laundry challenge and a four-week heating challenge, with a one-week overlap. The challenges were to:

1. Reduce weekly laundry cycles by half relative to the baseline. For example, if a household had reported an average of six laundry cycles per week, the challenge was to do just three laundry cycles per week.
2. Only heat to 18 degrees Celsius in living and sleeping areas, regardless of the reported baseline temperatures.

During the challenge period, all households were asked to complete weekly online surveys, through which they could also convey messages to the ELL researchers. In instances where households conveyed messages of despair, the ELL researchers would get in touch to offer verbal support or a bit of guidance. In general, however, the idea was not for the households to achieve the challenge goals no matter what, but to experiment with the challenges, and convey to the researchers what they viewed as possible (or impossible) throughout the process. Although the households were given autonomy in terms of how they wanted to approach the challenges, ENERGISE provided challenge kits to help participants implement less resource-intensive laundry and heating-related practices. These kits included items to support comfort and cleanliness, such as clothes racks to help organize and air clothes-in-use, as well as tips and tricks to challenge social expectations related to comfort and cleanliness, such as information about how the frequency of washing clothes has changed over time.

After seven weeks of experimentation, the ENERGISE researchers met with the households a third time in December 2018 using the same format as the second meeting, and asked participants to describe their experiences in experimenting with the challenges. Three months later, in March 2019, the households received a final survey to assess whether they had maintained the new patterns they had established during the ELLs.

In May 2019, a workshop was held with several of the ELL participants, local politicians from Roskilde Municipality and representatives of the laundry machine and building industry. At the workshop, the ELL researchers presented the results from the ELLs. The overall discussion topic for the workshop was: 'We need to reduce energy consumption—but where are the spaces for action and who is responsible?'

ENERGISE Living Lab Results

On average, the Danish households reduced their weekly laundry cycles by nearly 40 per cent, and indoor temperature by approximately 1 degree Celsius, with improvements both in terms of *realized* temperature changes and *experienced* comfort levels related to temperature. It is important to note that these changes came about because the households implemented new ways to keep their clothes clean and embraced new ideas about when clothes need to be washed, and implemented new ways of keeping their bodies warm. For example, almost all participating households adopted a sensory approach to determine when clothes needed to be washed (for example by examining for smells or stains), instead of automatically putting the clothes in the washing machine after wearing them one or two times. Three months later, most households had maintained the new patterns established during the ELL process.[2]

FRAMING AND STAGING SPACES IN ENERGISE LIVING LABS

As described above, the ELL process is composed of several steps, which can be framed as different negotiation spaces (Pedersen 2020) in a staging process. In the sections that follow, I elaborate on who and what staged the various spaces in the ELL process, how these spaces were configured, as well as the purpose of the spaces. Finally, I elaborate briefly on the types of knowledge generated through the spaces and how such knowledge travelled.

Who and What Framed and Staged ELL Spaces

First of all, in contrast to most participatory design staging processes (see examples in Chapters 1 and 2, this volume) the ELL researchers initiated and guided the ELL process; however, in line with participatory design staging processes, they did embrace the notion of the 'humble stage director' (a term coined by Pedersen 2020) who wants and needs to learn from the actors that she/he invites onto the stage. The researchers drew heavily on practice theory to frame the spaces and configure connections between those who were invited onto the stage—namely, the participating households, their practices and the material arrangements that co-configured those practices. Going beyond participants' observable behaviours, the researchers prioritized the underlying social, institutional and material conditions for those behaviours.

In line with Akrich (1992), the researchers viewed material arrangements as important actors that co-configured or co-scripted heating and laundry practices. However, in contrast to Sayes (2014), those who adopt a practice theoretical perspective would argue that material aspects of practices 'act' just as much as the meanings and competences that are part of (re)producing a practice. For example, laundry-related practices comprise different configuring elements ranging from socially shared ideas about cleanliness and presentability, to routines and competences related to how often clothes should be cleaned, to material arrangements concerned with how clothes are cleaned (often, by using soap and water and a washing machine). When a new material object (such as a clothing rack) is introduced into a particular practice or set of practices, it functions as an intermediary object (Vinck 2012) that makes it possible to flesh out the dynamic relationships among meanings, competences and materials involved in performing particular practices (such as those related to laundry). Although different actors were invited onto the stage at different times and performed in different space configurations, the ELL process was always guided by a practice theoretical framing. The primary aim for the entire process and for all of the spaces was to explore how energy demand can be reduced when it is understood as an outcome of the dynamic interplay of social practices.

The Configuration and Scale of ELL Spaces and the Performance of Staging

It can be argued that four different spaces were configured during the ELL process. Whereas participating households remained onstage throughout the process (arguably playing slightly different roles across the spaces), other types of actors (particularly material actors and policy actors) were invited onto the stage only at particular times. A sort of pre-space was configured during the initial work that structured the framing and design of the ELL process. To some extent, decisions about which actors were central to the ELL process were made *before* participating households were recruited. It had already been decided how many households to involve, which areas of consumption to target (heating and laundry) and why these areas of consumption were important based on a practice theoretical understanding of how to change energy and resource demand. This is somewhat different from other types of staging processes, where the framing of relevant spaces might emerge from a more participatory and politically navigating co-design process, as evident throughout the examples presented in this book.

The first space was configured at the beginning of the ELL process during the first meeting, the baseline data collection period and the first half of the second meeting. During this phase of the ELLs, the households, practices and routines were invited onto the stage for discussion and deliberation.

The second space was configured when the challenges were introduced during the second half of the second meeting, and lasted until the challenges ended. During this phase of the ELLs, the challenges as well as the challenge kits were invited onto the stage as new actors. The second space therefore comprised a new configuration of actors, resulting in existing practices and routines being challenged and reconfigured. For example, the clothing rack included in the laundry challenge kit helped participants organize and air the clothes they had worn, but did not yet want to wash. Tips about airing clothes instead of washing them also helped participants wear their clothes longer before washing them.

The third space was configured during the last steps of the ELLs when the participants, their revised practices and routines, as well as their prior practices and routines were invited onto the stage to deliberate about the changes that had (not) been made. During this phase of the ELLs, the old and new configurations of practices of keeping clean and comfortable met in deliberate confrontation, as participants described what had been easy and what had been difficult to change. This space continued to exist from the final visit until the follow-up survey as they continued to reflect on and experiment with their new practices.

The fourth and final space was configured and established during a follow-up workshop. This space was notably different, as local authority actors as well as industry actors were invited onto the stage to discuss spaces for action towards sustainable futures and who should be involved in them. What may be particularly interesting to note about this space is that the researchers/workshop facilitators played a particularly visible role as 'stage directors' insofar as we instructed the industry and local authority actors not only to be silent during the first part of the workshop, but to actively listen to the participants' experiences with the challenges. They could ask participants about their experiences, but they were instructed to neither reveal their own impressions of the process, nor provide recommendations or explanations representing their professional positions. During the second half of the workshop the 'silenced' actors were invited to reflect on different topics, taking what they had learned during the first half of the session as their point of departure. This was a deliberate decision designed to give actors who usually do not have a voice in policy or industry related discussions about (product) sustainability and change an opportunity to be heard.

The Travel of Knowledge through ELL Spaces

The overall problem framing remained the same throughout the ELL process. However, within each space, particular aspects were reframed when actors who met 'onstage' engaged in negotiation or contested each other. Participants' practices related to how to keep clothes presentable were reframed, as several participants began airing their clothes or spot cleaning to remove stains instead of immediately washing their clothes. Furthermore, and perhaps more interestingly, participants changed how they determined when an item needed to be washed, shifting from relating it solely to length of wear to the active use of senses, such as smell. In contrast, old practices related to heating were more difficult to modify. Reframing what was considered to be a comfortable living space proved very difficult as old practices to a larger extent prevailed; however, notions of a comfortable sleeping environment were reframed, as the challenges enabled many participants to experience better sleep in colder sleeping environments.

These reframings produced new (embodied) knowledge that travelled to subsequent spaces. In part, the knowledge travelled through the journal entries, as participants kept an active overview of how often they did laundry and why, as well as of how warm or cold their living spaces became. More interestingly, the knowledge generated through new experiences with how to keep clean and warm also travelled through the embodied new routines of, for instance, airing the clothes instead of washing them. In the third space, this knowledge played a significant role as participants drew the researchers' attention to their experiences with changes in their sleeping environments or changes made to the process of determining whether clothes were dirty or not. Noteably, these new experiences played a less significant role in the fourth space, where other aspects such as the accessibility and (in)comprehensibility of heating and laundry products and systems were more prominent. This had probably to do with the fact that new actors, such as politicians and representatives from private companies, were part of the fourth space, and the ELL participants wanted to bring problems related to products and access to systems to the attention of these new actors. This is a good example of how the actors that are invited onto a particular stage indirectly participate in framing the space, simply by representing a particular arena or framing from another space. Here, the invited representatives from the laundry machine industry, the building industry as well as the local government represented their companies or the municipality in a way that went beyond the scope of the workshop, simply by representing a sort of access to their respective fields. Although all the participants who were part of the workshop knew the overall goal of the workshop, it is impossible to avoid the fact that participating actors become spokespeople for the arenas or fields they represent in ways that can be difficult for the stager to control or even anticipate. This in itself is not a problem, but the 'stager' or 'stage director' needs to be aware of this potential when staging.

Overall, it became clear that information about alternative ways of maintaining cleanliness and comfort only translated existing practices when and if this information was accompanied by intermediary material objects (such as the clothes rack) *and* strategies for challenging existing socially shared norms (such as deliberating and challenging how many times a piece of clothing can be worn before it needs to be washed). In other words, when practices change or are reconfigured, the reframing is enacted and carried *across* meanings, skills and materials connected to those practices.

Although the knowledge and experiences that travel through practice changes might be longer lasting, knowledge about the project's outcome have also travelled outside the project and the immediate participants. Project results and recommendations have been transformed into newspaper articles, such as in Roskilde's local newspaper, endorsed by the chairman of the local climate and environmental council. The project design and results have also been published through popular science media (videnskab.dk) in addition to more traditional academic journals (see reference list for examples). Finally, the ELL approach and promising results have been transformed into a curriculum design for school projects that can be used at local schools in Roskilde.

A PRACTICE THEORETICAL CONTRIBUTION TO STAGING

A practice theoretical perspective might contribute to the concept of staging in several ways. First, because it draws on particular social ontologies, practice theory invites onto the stage a particular way of defining who and what is relevant to the process. Bearing in mind that the particular problem framings underlying traditional rational consumer choice models are (implicitly) prioritized in most political processes (Foulds et al. 2017), it is important to highlight that different problem framings, such as those from practice theoretical contributions, enable the configuration of new and refreshing spaces in which change can come about. In effect, using an alternative problem framing to co-define the space for change and action gives a voice to actors who are rarely consulted in traditional behavioural change programmes. Most notably, a practice theoretical problem framing invites onto the stage socially shared ideas about what a good and comfortable life is and its (potential) material composition. One the one hand, these shared ideas and related practices are given a voice, and on the other hand they are exposed to discussion and potentially challenged by other actors on the stage, including households, authorities, product developers and alternative (minority or prototypical) practices. In the following section, I highlight the main differences between practice-oriented staging and individual-oriented staging, paying special attention to methodological differences.

Practice-oriented Staging versus Individual-oriented Staging

As described in this chapter, a practice-oriented change process invites onto the stage a number of actors that are left behind the scenes in a more traditional individual-oriented behavioural change process, such as underlying ideas about the 'good' life, social norms related to these ideas, and practices that stubbornly reproduce (and are being reproduced by) these ideas. Likewise, a broader set of actors are invited onto the stage, as a practice perspective suggests that a broad range of societal actors are responsible for change, as opposed to only focusing on individual consumers. Thus, a practice-oriented change process opens up a much wider space for change than individual-oriented behavioural change processes.

Behavioural change processes most often focus on encouraging individuals to change their behaviours or attitudes by making other choices (Shove 2010). This approach seems to suggest that 'inefficient' behaviours can seamlessly be replaced with more 'efficient' behaviours by making individuals aware that they can make different choices (Labanca and Bertoldi 2018). For example, a behavioural change programme may focus on letting individuals know that washing clothes in 30-degree water consumes less energy than washing clothes in 40-degree water. This approach, however, typically black-boxes the frequency with which individuals wash their clothes and why. Practices, socially shared norms and discourses about cleanliness and presentability often (deliberately or unknowingly) remain behind the scenes. The stage is thus very narrow, but easily manageable, and the spotlight (or responsibility for action) shines solely on a single actor, who is expected to react to the information presented.

A practice-oriented change process, on the other hand, deliberately invites onto the stage the practices, socially shared norms, traditions and discourses related to doing laundry, so that both the frequency of washing clothes and the underlying reasons are surfaced, scrutinized and challenged to some extent. Meanings, competences and materials related to the practice of doing laundry unfold equally as relevant actors on the stage. The stage is much wider, and the spotlight shines on several actors who are viewed as playing equally important roles. This also means that the space for change becomes more complex and that interactions among multiple actors become important for the process.

Methodologically, a behavioural change programme typically is implemented during a single session wherein individuals are provided with information about 'best practices' and efficiency, and are given monetary incentives which, at best, result in limited levels of engagement. In a practice-oriented change process such as that implemented by the ELLs, however, an extended time-space for experimentation is provided for the participating actors to deliberate, scrutinize and challenge existing relationships and practices, such as how often clothes are washed and why, or the relationship between indoor temperature and notions of a comfortable and welcoming home. Through this process, norms that structure everyday life are deliberated and challenged. In the case presented in this chapter, participants who initially thought they could not reduce laundry cycles or indoor temperatures any more than they already had (and thus might not have tried had they been instructed to at a single session) were able to experiment with different laundry and heating-related practices and experience that change was in fact possible. In addition to the extended time-space, some of this was made possible—and importantly, visible—through the deliberate introduction of (new) materiality such as warm socks that heat the body instead of the room, or clothing racks that can be used to organize and air in-use clothing.

Thus, a practice-oriented 'play' is longer than an individual-oriented one, and the time and space required for practice-change programmes thus place more demands on the stage director and the actors, as well as the audience. A practice-oriented process may appear longer and more dynamic than an individual-focused process, as the change process cannot be ticked off as completed by the mere provision and distribution of information about more efficient behaviours.

CONCLUDING REMARKS: A STAGING-ORIENTED CONTRIBUTION TO PRACTICE THEORY

Just as practice theory has contributions to make to a concept of staging, the concept of staging has contributions to make to practice theory. In explicitly thinking about when and how different societal actors play their roles in a transition process, it makes sense to think about societal actors as actors who are invited on and off a stage for particular reasons. For instance, when laundry-related practices are challenged by introducing a clothing rack that enables people to air their clothes instead of washing them, a new dynamic evolves between actors on the stage, including households, clothes that need to be washed, clothes that are considered clean, ideas about cleanliness, competences related to cleaning clothes, and (intermediary) material objects, such as the clothing rack. In this case, the laundry machine is pushed off the stage, or at least the spotlight shifts from the laundry machine to the clothing rack, which makes all actors 'look' and interact differently. Although practice theory emphasizes that change affects all aspects of practices when a new element is introduced (be it material, meaning or competence related), a concept of staging makes it possible to explicitly explore the dynamic and, not least, gradual aspects of those changes.

Along these lines, and perhaps most interestingly, applying a concept of staging to the ENERGISE ELL process made me actively think about how old and new versions of practices intersect, confront each other and even negotiate through the people who maintain the practices, and that old and new practices (co-)exist during a change process. When actors such as households are invited to explicitly reflect about 'before and after' in terms of how they do laundry, and whether and how their ideas about cleanliness and presentability have changed, a space is created where both old and new versions of practices are verbalized and, most importantly, negotiated.

On a more meta-level, the concept of staging seems to offer a great metaphor for explaining the role of theories and problem framings in knowledge production. Thinking about an experimental intervention process as something that is actively staged with a scene, props, actors and a manuscript enables an awareness of how theories provide 'worldviews' that to a large extent define who should be on the stage (actors) and how they should interact; how theories treat (and co-perform) the dynamic relationships among societal actors, their roles in sustainability transitions, and their levels of responsibility; and how theories (co-)define aspects of power and potential spaces of action.

ACKNOWLEDGEMENTS

I would like to thank all of the ENERGISE team members and partners who collaborated on this project, as well as all of the household members who participated in the Living Labs. The European Network for Research, Good Practice and Innovation for Sustainable Energy (2016–2019) project (ENERGISE; http://energise-project.eu) was coordinated by the National University of Ireland, Galway, and funded by the European Union's Horizon 2020 programme (GA No. 727642). I would also like to thank the editors of this book for the opportunity to contribute to the book and for the feedback on my chapter.

NOTES

1. If you want to establish your own ELL, you can find the design manual here: http://energise-project.eu/livinglabs (accessed 6 August 2020).
2. For more details on types of practice changes, please consult the report on general ENERGISE results, http://energise-project.eu/sites/default/files/content/ENERGISE_D5%202_260919_Final.pdf, and the Danish results, http://www.energise-project.eu/sites/default/files/content/ENERGISE%20pixibog_endelig_interaktiv.pdf (both accessed 6 August 2020).

REFERENCES

Akrich, M. (1992), 'The de-scription of technical objects', in W.E. Bijker and J. Law (eds), *Shaping Technology / Buiding Society: Studies in Sociotechnical Change*, Cambridge, MA: MIT Press, pp. 205–24.

EC (European Commission) (2019), *Report from the Commission to the European Parliament and the Council 2018. Assessment of the progress made by Member States towards the national energy efficiency targets for 2020 and towards the implementation of the Energy Efficiency Directive as required by Article 24(3) of the Energy Efficiency Directive 2012/27/EU*, Brussels: European Commission.

EEA (European Environment Agency) (2013), *Technical Report No 5/2013: Achieving Energy Efficiency through Behaviour Change: What Does It Take?*, Copenhagen: European Environment Agency.

Foulds, C. and T.H. Christensen (2016), 'Funding pathways to a low-carbon transition', *Nature Energy*, **1**, 16087.

Foulds, C., R. Robison, L. Balint and G. Sonetti (2017), *Headline Reflections—SHAPE ENERGY Call for Evidence*, Cambridge: SHAPE ENERGY, accessed 14 February 2020 at https://shapeenergy.eu/wp-content/uploads/2017/07/Call-for-evidence_reflections.pdf.

IPCC (2014), *Climate Change 2014: Synthesis Report. Contribution of Working Groups I, II and III to the Fifth Assessment Report of the Intergovernmental Panel on Climate Change*, ed. R.K. Pachauri and L.A. Meyer, Geneva: IPCC.

IPCC (2018), *Global Warming of 1.5°C: An IPCC Special Report on the Impacts of Global Warming of 1.5°C above Pre-industrial Levels and Related Global*

Greenhouse Gas Emission Pathways, in the Context of Strengthening the Global Response to the Threat of Climate Change, Sustainable Development, and Efforts to Eradicate Poverty, ed. V. Masson-Delmotte et al., Geneva: IPCC.
Jensen, C., G. Goggins, I. Røpke and F. Fahy (2019), 'Achieving sustainability transitions in residential energy use across Europe: The importance of problem framings', *Energy Policy*, **133**, 110927.
Jensen, C., G. Goggins, F. Fahy, E. Grealis, E. Vadovics, A. Ganus and H. Rau (2018), 'Towards a practice-theoretical classification of sustainable energy consumption initiatives: Insights from social scientific energy research in 30 European countries', *Energy Research and Social Science*, **45**, 297–306.
Laakso, S., K. Matschoss and E. Heiskanen (2019), 'Online tools and user community for scaling up ENERGISE Living Labs', ENERGISE—European Network for Research, Good Practice and Innovation for Sustainable Energy, Deliverable No. 3.6.
Labanca, N. and P. Bertoldi (2018), 'Beyond energy efficiency and individual behaviours: Policy insights from social practice theories', *Energy Policy*, **115**, 494–502.
Pedersen, S. (2020), 'Staging negotiation spaces: A co-design framework', *Design Studies*, **68**, 58–81.
Sayes, E. (2014), 'Actor–network theory and methodology: Just what does it mean to say that nonhumans have agency?' *Social Studies of Science*, **44** (1), 134–49.
Shatzki, T. (2002), *The Site of the Social: A Philosophical Account of the Constitution of Social Life and Change*, Philadelphia: Penn State University Press.
Shove, E. (2010), 'Beyond the ABC: Climate change policy and theories of social change', *Journal of Environment and Planning*, **42**, 1273–85.
Shove, E. (2017), 'What is wrong with energy efficiency?', *Building Research & Information*, **46** (7), 779–89.
Shove, E. and G. Walker (2014), 'What is energy for? Social practice and energy demand', *Theory, Culture and Society,* **31**(5), 41–58.
Shove, E., M. Pantzar, M. Watson (2012), *The dynamics of Social Practice – Everyday Life and How it Changes*, Thousands Oaks, CA: Sage Publications.
Sovacool, B.K. (2014), 'What are we doing here? Analyzing fifteen years of energy scholarship and proposing a social science research agenda', *Energy Research and Social Science*, **1**, 1–29.
Spurling, N., A. McMeekin, E. Shove, D. Southerton and D. Welch (2013), 'Interventions in practice: Re-framing policy approaches to consumer behaviour', Sustainable Practices Research Group. Accessed 6 August 2020 at: http://eprints.lancs.ac.uk/85608/.
Vinck, D. (2012), 'Accessing material culture by following intermediary objects', in L. Naidoo (ed.), *An Ethnography of Global Landscapes and Corridors*, Rijeka, Croatia: InTech, pp. 89–108.

13. Storytelling urban nature: situated intervention as environmental theatre

Ask Greve Johansen and Hanne Lindegaard

This chapter's focus is on two urban nature strategy workshops held with a diverse group of professionals working with municipal planning in Copenhagen. Urban greening – the rearrangement of urban space around an imaginary of nature – has become a ubiquitous principle for renewal and redesign of urban environments across the globe (Angelo 2019). In the City of Copenhagen, new policies for dealing with urban nature prompts planners and administrators to transform practice to facilitate 'greening'. This chapter describes our attempt to participate in and shape the process of stabilizing action around a hybrid conceptualization of 'urban nature'. We use the staging of human and non-human actors to inquire into the stabilization of planning practice as a form of situated intervention (Zuiderent-Jerak 2015), helping to sort and articulate urban nature attachments in a setting characterized by an abundance of normativities. This process reveals that the material-semiotic quality of urban nature relevant to municipal actors is grounded primarily in documents and suggests further storytelling to connect bureaucratic practice with concern for urban nature.

NATURE IN CITIES

Scholarship has emphasized urban ecologies and urban nature as distinct aspects in the production of inequality (Anguelovski et al. 2019; Bryson 2013; Cucca 2012; Gandy 2004; Heynen 2003, 2006; Swyngedouw 2006), but has also highlighted how urban nature in various forms potentially reconfigures socio-material relations between actors such as humans, animals, plants, water, earth and buildings with possibly positive implications (Angelo 2019; Bryson 2013; Classens 2015; Dorst et al. 2019; Hinchliffe and Whatmore 2006). Many conceptualizations of urban nature are rooted in a historical understanding of 'nature as something that resides outside of social relations' (Gandy 2006: 70). Over recent decades, urban spaces characterized by hybrid or overlapping forms have sought to challenge this nature–society dualism through urban design (Loughran 2016). Accordingly, (hybrid) rearrangements of urban nature have implications for which values are attributed to urban nature arrangements and whom such values might benefit. While 'green is good' might indeed be a ubiquitous notion, the specific arrangement of urban nature is a matter of great contestation among citizens, politicians, and professionals in the municipal bureaucracy.

The administration of the City of Copenhagen takes its point of departure in a steadily growing green agenda among citizens and elected politicians. Attention is focused mainly on how to change, adapt, and reformulate the administration's practices to align with emergent visions of the 'good' life in the City. Complementing the different normative attachments to urban nature articulated by citizens, politicians, and urban professionals are intra-organizational differences expressing a range of viewpoints, concerns, and commitments to particular areas or solutions. As might be said of planning practice in general (Healey 2004), concepts, ideas, and actor positions are constantly shifting in the City of Copenhagen's planning practice. Our intervention related to 'urban nature' as a new discourse articulated in the document 'Urban Nature in Copenhagen. Strategy 2015–2025', which was ratified by the City of Copenhagen in 2015 (Københavns Kommune 2015a, 2015b). The City of Copenhagen's Technical and Environmental Administration (TEA) generally defines 'urban nature' as encompassing all non-human life within cities, but as implicitly limited to a more tangible subset of animals and plants. Urban nature is described as a continuum ranging from green roofs, façades, and tree-lined streets to larger green areas in proximity to urban districts. The approval of the urban nature strategy signals increased political attention to how 'nature' is enacted in urban space. As a policy objective, the Copenhagen urban nature strategy emphasizes the benefits of urban nature in terms of recreational opportunities and related implications for health, climate regulation, decreased noise and air pollution, and protection against biodiversity loss. Articulating urban nature as a multifunctional good, the overall mandate in the strategy is to create 'more and better' urban nature. The abundance of 'goods' (that is, normativities) implied by the local notion of urban nature renders it accessible to the sorting and rearticulation work implied in Teun Zuiderent-Jerak's (2015) notion of 'situated intervention'.

SITUATED INTERVENTION

This chapter's focus is on two urban nature strategy workshops held in the spring of 2019 and the associated staging process. The first author had been working as apprentice and ethnographer in the TEA since January 2017 in a department tasked with developing new strategic initiatives, with the TEA as hosts and sponsors of his doctoral work. This situation formed a frame for regular participation and ethnographic inquiry (Lave 2011). Whereas

Johansen's apprentice role was welcomed, he was also expected to contribute to the TEA by conducting practice-relevant research. His challenge was to find a productive path; rather than directly adopting the problems and narratives articulated by management and other actors, he had to find a way to contribute expertise while remaining committed to what Zuiderent-Jerak (2015: 180) called the 'specifically problematic'. As has long been argued in the science and technology studies (STS) literature, matters of knowing and expertise cannot (and should not) be viewed as separate from matters of normativity and politics (see, for example, Law 2017). How we structured our roles and expertise in our relations with the TEA was also about the normativities implied in these relations. We were reluctant to construct what Zuiderent-Jerak (2007) called a 'deficit model' of normativity, where the role of the researcher is to introduce a normative stance to a situation presumed to lack normativity. In addition, we did not want to presume that the situation was settled or unproblematic or to adopt an agenda laid out by TEA management. Instead, our approach resembled Zuiderent-Jerak's (2007: 316) notion of 'the normative surfeit mode … refiguring the role of the STS researcher as one of dealing with the overflowing normativities encountered in a setting one is engaging with'. At the core of Zuiderent-Jerak's notion, then, is a shift from using engagement as an (external) resource to be drawn upon to seeing interventions as part of 'the context of particular, concrete circumstances' (Suchman 1987: viii, as cited in Zuiderent-Jerak 2015). Intervention is thus less a tool and more a part of the situation on which to reflect and construct knowledge from, and objectivity is sought not through detachment, but by being somewhere in particular (Haraway 1988).

ENVIRONMENTAL THEATRE

During his apprenticeship, Johansen had identified urban nature as a somewhat challenging topic for the TEA in cross-cutting projects, because of the multiplicity of tangible and abstract ideas about what 'urban nature' was. Through the workshop format, we sought to stage the discovery and possibly sorting out of this multiplicity as dramatic events. Inspired by work in the codesign tradition, we were interested in exploring how collaborative design practices might open the research practice towards rehearsing other futures (Halse et al. 2010) and extend available forms of political mediation and representation (Binder et al. 2015). To clarify the staging work involved in the two workshops and their preparation, we structure this description around a theatrical metaphor (Binder 2014; Brandt and Grunnet 2000). This enables us to emphasize the dynamics associated with setting the stage, (not) following the script and facilitating and participating in the two workshops, and how tensions can open a field to analysis. We use theatrical metaphors to describe the preparation, execution, and interpretation of the workshops and use the staging concept as an overall metaphor for these phases. During the preparatory phase, we worked on the script and identified relevant props and actors[1] for the workshops. We describe the workshops as an unfolding drama and explain how we provided participants with cues as we attempted to historicize their practice. The theatrical metaphor points to relevant components of intervention situations, and through the concept of staging, demonstrates that the situation is not 'natural' but exists as a result of staging work.

The use of the theatre metaphor is limited by the wide range of events and performances that may be considered theatre (see also Chapter 15, this volume). An important point in our findings emerges from the contrast between the *planned* workshops and the *actual* workshop situations as spaces of distributed and ambiguous power and agency. The diversion from our prepared script described below necessitates a vocabulary capable of comparing different genres of staging. In this regard, Schechner's (1968) concept of 'environmental theatre', which was developed to capture contemporary theatre productions' efforts to challenge established dramatic formalisms (Aronson 2010; Schechner 1968), is useful for characterizing situations where the distribution of roles among the director, performers, and audience members is underdetermined and fluid. The environmental theatre concept thus provides a framework for thinking about intervention situations with minimal assumptions about spatial composition, role distribution, and agenda.

Drawing on Goffman (1961, 1963), Schechner aimed to produce a relational, rather than a descriptive definition of theatre. The analytical utility of environmental theatre is to emphasize the theatre as constituting an environment for interaction rather than as a set of fixed roles. While retaining a conceptual distinction between performance and audience, Schechner (1968: 64) emphasized the variable relationship between script and performance: 'The text is a map with many possible routes. You push, pull, explore, exploit. You decide where you want to go. Rehearsals may take you elsewhere. Almost surely you will not go where the playwright intended.' Adopting this metaphor, we show how our role as propositioning facilitators (Binder et al. 2015) was negotiated in response to a situated intervention, rather than deployed strategically at the intervention's outset. In the sections that follow, we describe the process of preparing and conducting two workshops that form the empirical focus of the chapter.

STAGING INTERVENTIONS

During Johansen's ongoing fieldwork, he had become interested in exploring the multiplicity of urban nature within the TEA. Building on a hunch, the workshops were developed as part of an ongoing abductive investigation

(Tavory and Timmermans 2014; Timmermans and Tavory 2012). In other words, we intended the workshops to lead to what Donald Schön (1984) called double-loop learning—exploring existing ideas and practices about urban nature while simultaneously challenging existing problematizations thereof within the TEA. The workshop format presented an opportunity for us to stage and unfold this process; in theatrical language, this was the drama that we expected to be performed.

Following the notion of situated intervention, we were wary of seeing our hunches as more than that. Looking for the specifically problematic (Zuiderent-Jerak 2015) required us, on the one hand, not to set aside our curiosity and normative attachments and, on the other hand, not to think of this 'baggage' as anything other than present in the situation we were exploring. However, a need to make the purpose of the workshops clear required us to frame the interaction for participants. Thus, we communicated the overall objectives of the workshops as (a) to expand the conceptual space of 'urban nature', and (b) to prototype planning objects that might value urban nature differently.

We wanted to connect open-ended exploration with a more tangible outcome of tools that would better encompass the different conceptualizations of urban nature in the TEA. We referred to this intention as creating 'stronger management tools' in the illustrated two-page PDF document we sent with the invitation. This objective was to be achieved through two workshops one week apart attended by people representing different departments and different organizational levels within the TEA. The first workshop was intended to unpack different meanings of urban nature in TEA practice and from there to build a 'product specification' for a more potent management tool. The second workshop would be devoted to developing prototypes for new planning objects to value urban nature. The schedule and our approach to developing and presenting the workshop objectives formed the script for the intervention.

The workshops were held just after the TEA had undergone a series of stressful events concerning new budgets, organizational changes, a hiring freeze, etc. These changes left few TEA staff members available to participate in the workshops. To encourage participation, we decided that the workshops would take place onsite and emphasized them as being part of a strategic decision by TEA management. We aimed to design our workshops in a manner that would appear valuable to the participants involved and to make this value apparent in our invitation. We identified relevant invitees through informal conversation with persons we knew to be involved in urban nature work. We specified that we aimed to include people from across the TEA and encouraged recipients to forward the invitation to relevant colleagues in order to acknowledge different forms of urban nature expertise in the TEA.

Figure 13.1 *A table with props from the workshop*

In addition to making a schedule, preparing sticky notes, procuring pens and index cards in different colours, and developing a set of exercises, we collected objects (props), to represent different forms of urban nature, including: Japanese knotweed, *fallopia japonica* (an invasive species present in Copenhagen as well as in many other urban settings); a frame from a beehive holding honeycomb; a sample of local lake water containing various plant material, cigarette butts, etc.; hides from a fox and a badger; a plastic insect net; a full dog poop bag; and a printout of 'Urban Nature in Copenhagen Strategy plan 2015–2025'. We also included pictures of multiple and different locations throughout the City to provide a current snapshot of the urban landscape (a cloudburst mitigation roundabout, seagulls eating bread and breakfast cereals, a municipal park in bloom, alpacas grazing on municipal land, an 'urban grove' at the city hall square, a green roof covered with sedum, flower pots mounted on a building façade, and barely visible grasses on top of an office building, etc). We thought these objects and pictures would resonate well with the workshop participants and might provoke or elicit comments (see Figure 13.1).

We planned a role distribution where Johansen would act as the main facilitator, and Lindegaard would observe and interject, but play a more passive role. We sought to align the workshops with a mandate from TEA management, to both secure participation and create a connection where knowledge generated in the workshops would inform TEA practice. In doing so, we implied that the problems discussed in the workshops would be TEA's problems and that we as facilitators would be able to help address them, what Michel Callon (1986) referred to as problematization. Invoking management to justify our presence, we sought to gain legitimacy at the risk of inheriting any mistrust of management that would be present.

Our configuration of the space for the workshops was an attempt to re-problematize urban nature as it appeared to planning and management practice in the TEA. The work involved in this process focused on obtaining an understanding of 'the situation' at the TEA, both regarding current problematizations of urban nature and organizational factors that might influence the process. It also included the production of two texts, the script for the workshop, and the invitation to send to potential human participants. Finally, we collected and displayed materials we expected might emerge as relevant non-human actors during the workshops.

THE DRAMA UNFOLDS

In the first workshop, 11 participants were gathered around one table. As they represented different parts of the TEA, many of them did not know either of us or each other particularly well. After a short introduction, Johansen presented the agenda for the two workshops. A couple of participants raised their hands. One participant questioned our intention to create stronger planning objects, as they felt the existing tools at their disposal (referring and pointing to the printout of the 'Urban Nature in Copenhagen. Strategy 2015–2025' document) were sufficient. Several other participants agreed with this point, saying that 'there was no need to develop strong planning tools', and 'further quantification of urban nature would not improve their capacity to further the urban nature agenda'. This was an immediate challenge for us. We had begun the workshop with a well-planned agenda and a clear idea of how the time at our disposal would proceed. Before we had got anywhere, Johansen, cast in the role of facilitator, had become the main antagonist in the unfolding story. The dominant sentiment in the situation was that our prepared agenda was not what was needed, yet it was unclear what an alternative path forward would be.

We saw no better way forward than to follow our planned agenda with some delay. We asked participants to select one of the pictures or one of the objects we had brought to the workshop that mattered to them in their work with urban nature. After selecting their objects, participants split into smaller groups of three and sat together at smaller tables to share why they had chosen their objects. Choosing the objects was a rather straightforward process. Johansen took a more passive role, and we both circulated the groups to pick up on the conversations. Once objects were introduced into the conversations, the rather large group was split into smaller ones, and the story's antagonist had taken a more passive role, the mood changed, and the dialogue shifted to a more constructive mode.

We reconvened, and the participants reported on their group discussions. One important point was that 'urban nature' was something profoundly different from 'nature'. The participants emphasized that 'urban nature' was a *representation* rather than an *instance* of 'nature' in an urban setting. At the risk of oversimplification, 'urban nature' was deemed effective because it simplified and represented nature. The agency of the concept of 'urban nature' amounted to how well it 'translated' notions of nature into ways of managing and planning urban space.

Participants highlighted important differences between the City's planning and management functions regarding what forms of urban nature were relevant and which instruments they would use. They often experienced tensions between their desire to develop ideal solutions and the reality of economic constraints. Moreover, networks needed to be built within the organization to secure a mandate to work for 'more and better' urban nature. We collectively concluded that the next workshop should focus on sharing 'positive' stories that would illustrate the work already being done, rather than developing new governance objects to value/quantify urban nature. At this point, we had re-established some trust with the participants and regained our bearings. Working with stories was meaningful to most of the participants and we had learned, importantly, that the 'Urban Nature in Copenhagen. Strategy 2015–2025' document was referred to as a solid and important basis for 'urban nature' work.

However, we were also left with some unanswered questions. Our idea of 'stronger management tools' had met prompt resistance from the participants. We were unsure whether we had underestimated the effect of having management present, in terms of stressing that the workshops were the result of a management decision, and by including a manager as a participant in the workshop. On the other hand, the introduction of our curated props/non-human actors to the conversation had not only prompted participants to explicate how urban nature mattered in their work but had also made clear how 'urban nature' mattered to participants as representation rather than instance of nature. The sample of lake water, for example, was relevant because it represented improved water quality and the TEA's efforts that had produced this improvement. Workshop participants did not consider the sample itself 'nature'. Furthermore, the introduction of the props made a visible and audible difference in the conversation among participants. Shifting 'urban nature' from a strategic entity to a tangible object seemingly enabled participation on a more equal footing and defused most of the tension produced in the awkward beginning. Still, we wondered whether stating the intended outcome of the workshops made participants feel like they had limited influence over the process. By introducing props to the conversation, we intended to reveal indeterminacy regarding what counts as urban nature in TEA practice, prompting a reflexive conversation. As for what urban nature *was,* this did not have the intended effect. Instead, however, participants were eager to explore how (urban) nature could be *represented* in different ways.

Enrolment, Strategy, Critique

In the second workshop, two participants had participated in the first workshop, and three were new. The workshop began with the second author presenting an account of a previous process of stabilization of sewers in the City of Copenhagen in the mid-nineteenth century. The presentation was a case story focused on the problematization of public health and the discursive elements, networks, and problem spaces relevant to the precursor to the TEA some 150 years earlier. By historicizing the practice of stabilizing planning and management practices around another matter of concern, we hoped that participants would be better equipped to see their current work related to urban nature as a process of constructing governable objects. By comparing public health initiatives in the mid-nineteenth century with current urban nature initiatives and relating participants' current practices to those of early medical professionals and engineers, Lindegaard's presentation cued participants in on the reproblematization of urban nature.

We used a whiteboard to illustrate the story and supplemented it with contemporary themes and issues regarding urban nature as the discussion progressed. We used postcard-sized cardboard in two different colours to frame the conversation. Yellow cards represented more or less defined collective (mostly human) actors such as organizational units within the TEA, children, or developers. Pink cards represented non-human actors such as architecture policy, Roundup/glyphosate, or evidence. We had prepared a stack of cards for the workshop based on what we anticipated might be relevant to the discussion, and together with the other participants, filled out blank cards as the conversation progressed. The discussion centred around how to identify other actors and define their roles, what Michel Callon termed interessement and enrolment (Callon 1986).

An important theme was configuring the role of the citizen in relation to the strategy for urban nature. 'Citizens' in the abstract figured in a role as holders of sometimes unrealistic expectations, but also as a catalyst for change within the system, providing an additional and perhaps more direct mandate for 'more and better' urban nature. Managing the expectations of citizens was emphasized more at the beginning of the discussion; towards the end, citizens were primarily discussed as an additional political resource. In both cases, the task of the bureaucrat was to manage the citizen figure, thus locking the citizen actor into a specific role. In terms of enrolment, a major theme was how the notion of public health could strengthen the relations between established actors and expand the network. Among the suggestions presented was a collaboration with the City's Health and Care Administration to promote the health benefits of green spaces, and specifically working towards establishing therapeutic gardens in municipally managed cemeteries.

Our script for the second workshop deviated from our initial plan to prototype planning objects, and we allowed the discussion to progress along a more loosely defined path. Instead of a script for action, we facilitated the workshop through the historical case and the cards, and the result was that storytelling entered the stage. As the discussion unfolded, we transitioned from participating in the open conversation to leading it, as we wanted the workshop to end with a solid conclusion. During the first workshop, the conversation had moved from Johansen cast as the antagonist to working out a mandate for how to proceed, to identifying participants' matters of concern as they interacted with the 'props' (that is, the material objects we had brought for the workshop). We had decided to share stories that emphasized the work already being done and to illustrate some of the inherent tensions in urban nature work. Historicizing the overall practice of stabilizing planning and management practice in the second workshop was useful in this regard. Participants were able to translate the concrete historical example to their current practice, thus offering a new interpretive frame for reflecting on current urban nature work. The second workshop thus played out more collaboratively.

A substantial outcome of the second workshop was that our understanding of what role stories could play was made explicit: stories could serve as alternative accounts of what was achieved through bureaucratic work. The point of this was not as much to highlight completed projects, but rather to make the *ongoing* work visible to management, city councillors, and citizens. Two critical factors contributed to a smoother experience in the second workshop. First, the second workshop was smaller than the first, making it easier to manage a collective conversation. Second, no participants had managing roles in our second workshop. In that sense, there was less at stake and more room for an open discussion. The historicizing of current practice worked well. The effort we took before the workshops to learn how urban nature was discussed and documented, and how work was done in the organization helped us select a relevant case and enabled us to discuss it in a way that was relevant to participants.

THINKING THROUGH THEATRE

What can we learn from this process? During the workshops, the inclusion of props and our prepared historicization of planning practice facilitated the articulation of material sensitivity in our conversations. By repeatedly reframing the situation, we collectively arrived at a new understanding of what was needed to strengthen the TEA's strategic work on urban nature. Rather than establishing better ways of accounting for the value of urban nature, our shared task was to make the value of urban nature apparent by crafting stories that related the concept of urban nature (as a representation) to struggles and achievements that mattered in bureaucratic work to care for nature in urban space.

We can describe this following a dramatic structure over three acts. The first act involved the initial problematization of urban nature and, in a broader sense, the fieldwork preceding the workshops wherein a researcher entered the organizational practice and wondered exactly what urban nature meant and what one could do to intervene in practice. In the second act, a range of actors gathered on stage to design new ways of valuing urban nature. However, much to the two researchers' surprise, the actors rejected the script, and a new way forward had to be found quickly (the dramatic climax). In the third act, the actors collectively renegotiated the script through ongoing improvisation and arrived at a new way of understanding what was at stake and what should be done. By the end of the performance, the roles of participants, and the beliefs they had held when they had first stepped onto the stage had changed substantially.

When the script is unexpectedly rejected, it might be difficult to communicate events clearly relying on conventions in theatrical performance. The staging perspective described by Pedersen et al. (Chapter 2, this volume) concerns how diverse actors, material objects, and meaning are or can be drawn together in a common space to enable and facilitate interaction and co-creation, while accounting for contingency, temporality, and distributed agency. Importing a set of assumptions about the world championed in ANT lends a vocabulary to describing and understanding innovative situations as non-linear. We share the relational and material-semiotic affinities implied in the staging perspective, but in this account, we emphasize the environmental theatre as a form of discovery at the expense of creation.

Instead of abandoning the theatrical metaphor when it falls short, we propose further qualifying the type of theatre involved. The staging concept implies active, overt, and strategic action directed at an outcome (Pedersen et al., Chapter 2, this volume). In some situations, the actors initiating the staging might be deliberately unclear about the outcome. Returning to Schechner's notion of the script as a map with many paths to take, staging can be thought of as configuring and performing a navigation through a situation. Likewise, Schechner's environmental theatre concept is useful for challenging assumptions about the boundaries of the stage and, importantly, in this case, the distribution of roles among participants. This also implies that power might not be held only (or even mostly) by those who initiate the staging. Whether a script is absent or present might also matter less than the gathering of actors in a common space. The use of the concept for thinking about staging is to challenge preconceptions about site, text, and agency, which leads back to the notion of navigating through 'situated interventions' rather than formulating programmatic frameworks to outline 'how it is done'.

As daily life in a bureaucratic institution is highly discursive, the most important material objects may often be texts (Mangset and Asdal 2019). This point leads us to reconsider the relevancy of our prepared objects. As most participants were more interested in narrative representations of urban nature than for example plants and critters on their own behalf, they leaned towards reports and stories at the expense of dog poop bags and invasive plants. On the other hand, the inclusion of objects foreign to the office may have prompted the change in mood crucial for our recovery from the initial crisis.

The rejection of our prepared agenda in the first workshop can at least partly be explained with reference to Zuiderent-Jerak's notion of the specifically problematic as a core concern in the context of situated intervention. By stating the intended outcome of the workshops (that is, stronger management tools) in the invitation and the agenda, we provided a solution before we had identified the actual problem, leaving too little space for situation-specific problems to arise.

However, it would have been difficult to recruit participants to a workshop without a clear goal. In other words, an interesting tension exists between open-ended processes and the legibility of staging in cases where time is sparse, and management cannot be relied upon to make participation mandatory. On the other hand, the introduction of an antagonist who proposed a solution that was definitely *not* the way to go might have prompted participants to reflect on what should be done instead. The irritation caused by our staging of the workshops thus prompted the articulation of stories as a tool to connect bureaucratic practice with concern for urban nature, helping participants navigate a 'normatively abundant' (Zuiderent-Jerak 2007) matter of concern.

Similarly, the introduction of objects to the process produced a range of effects. As argued by Dominique Vinck (2012), professional practice is often heavily invested in intermediary objects. By intervening through objects, we sought to mirror the material aspect of practice, thereby enabling temporary intermediary objects to emerge. The lake water sample briefly mentioned above was revealed as a representation of the collective efforts to purify

Copenhagen's waterways and retrofit its sewers. We brought it to the workshop to exemplify the overfeeding of waterfowl and nutrient surplus, but it was effectively translated into a story about successful water management initiatives over the last 25 years. Given the status of the workshops as staged and temporary spaces, it is important not to put these temporary intermediary objects on equal footing with more established objects circulating in professional practice. However, in a 'local' sense, the introduction of tangible objects to the discussion clarified the matter at hand for participants and made it visible to us in our capacity as researchers (Vinck 2012).

CONCLUSION

The workshops articulated the concept of 'urban nature' in TEA practice in multiple forms. They revealed that municipal actors think of urban nature as a necessary simplification, enabling a relatively well-established agenda to promote environment and wellbeing in urban space through representations of nature. As 'designed' events in the context of sustained participant observation, the two workshops led us to re-evaluate organizational actors' positions in urban nature planning and revealed how we, the subject matter, and other actors, could be shifted into new positions, enabling new agencies.

Rather than resulting in a redefined notion of urban nature, the process revealed what was specifically problematic about urban nature for participating actors. Participants had little interest in defining urban nature or challenging the ubiquitous notion that green is good. Referring to the physical copy of the approved strategy, they regarded the matter as settled and resisted our attempts at redefinition by crafting new tools. However, working with the hybrid urban nature concept was experienced as dispersed initiatives, suggesting the need to craft stories to traverse the organization and connect bureaucratic practice with citizens' concerns and politically given objectives.

Our experience outlined above leads us to two observations: First, that the stage was not empty as we arrived. Despite our efforts to frame the interaction, we entered an already populated world, encountered an already present understanding of the subject matter, and the semiotic materiality of the office setting. Second, and following from this, specifying the type of theatre unfolding and how its form affects the action improves the analytical utility of the theatre metaphor. While our engagement with performance literature here is limited, further cross-pollination from more recent performance studies literature might enrich the staging vocabulary further.

NOTE

1. We use the term 'prop' to denote latent non-human actors in theatrical terms. In actor–network theory (ANT), both humans and non-humans must be brought to the stage to be considered actors (see for example, Latour 2005).

REFERENCES

Angelo, H. (2019), 'The greening imaginary: Urbanized nature in Germany's Ruhr region', *Theory and Society*, **48**, 645–69.

Anguelovski, I., J.J. Connolly, M. Garcia-Lamarca, H. Cole and H. Pearsall (2019), 'New scholarly pathways on green gentrification: What does the urban "green turn" mean and where is it going?', *Progress in Human Geography*, **43** (6), 1064–86.

Aronson, A. (2010), 'Environmental theatre', in *The Oxford Companion to Theatre and Performance*, accessed 8 July 2019 at http://www.oxfordreference.com/view/10.1093/acref/9780199574193.001.0001/acref-9780199574193-e-1236.

Binder, T. (2014), 'Design laboratories as everyday theater: Encountering the possible', in *Research Network for Design Anthropology : Ethnographies of the Possible*, accessed 13 December 2019 at https://adk.elsevierpure.com/da/publications/design-laboratories-as-everyday-theater-encountering-the-possible.

Binder, T., E. Brandt, P. Ehn and J. Halse (2015), 'Democratic design experiments: Between parliament and laboratory', *CoDesign*, **11** (3–4), 152–65.

Brandt, E. and C. Grunnet (2000), 'Evoking the future: Drama and props in user centered design', in *Proceedings of Participatory Design Conference (PDC 2000)*, Palo Alto, CA: Computer Professionals for Social Responsibility (CPSR), pp. 11–20.

Bryson, J. (2013), 'The nature of gentrification', *Geography Compass*, **7** (8), 578–87.

Callon, M. (1986), 'Some elements of a sociology of translation: Domestication of the scallops and the fishermen of St Brieuc Bay', in J. Law (ed), *Power, Action and Belief: A New Sociology of Knowledge?*, London: Routledge, pp. 196–223.

Classens, M. (2015), 'The nature of urban gardens: Toward a political ecology of urban agriculture', *Agriculture and Human Values*, **32** (2), 229–39.

Cucca, R. (2012), 'The unexpected consequences of sustainability. Green cities between innovation and ecogentrification', *Sociologica*, **6** (2). DOI: 10.2383/38269.

Dorst, H., A. van der Jagt, R. Raven and H. Runhaar (2019), 'Urban greening through nature-based solutions: Key characteristics of an emerging concept', *Sustainable Cities and Society*, **49**, 101620.

Gandy, M. (2004), 'Rethinking urban metabolism: Water, space and the modern city', *City*, **8** (3), 363–79.

Gandy, M. (2006), 'Urban nature and the ecological imaginary', in N. Heynen, M. Kaika and E. Swyngedouw (eds), *In the Nature of Cities Urban Political Ecology and the Politics of Urban Metabolism*, London; New York: Routledge, accessed 22 May 2019 at https://ezproxy.aub.edu.lb/login?url=https://www.taylorfrancis.com/books/9781134206476.

Goffman, E. (1961), *Encounters: Two Studies in the Sociology of Interaction*, Indianapolis, IN: Bobbs-Merrill.

Goffman, E. (1963), *Behavior in Public Places: Notes on the Social Organization of Gatherings*, new edition, New York: The Free Press.

Halse, J., E. Brandt, B. Clark and T. Binder (eds.) (2010), *Rehearsing the Future*, Copenhagen: The Danish Design School Press.

Haraway, D. (1988), 'Situated knowledges: The science question in feminism and the privilege of partial perspective', *Feminist Studies*, **14** (3), 575–99.

Healey, P. (2004), 'Creativity and urban governance', *Policy Studies*, **25** (2), 87–102.

Heynen, N.C. (2003), 'The scalar production of injustice within the urban forest', *Antipode*, **35** (5), 980–98.

Heynen, N.C. (2006), 'Green urban political ecologies: Toward a better understanding of inner-city environmental change', *Environment and Planning A: Economy and Space*, **38** (3), 499–516.

Hinchliffe, S. and S. Whatmore (2006), 'Living cities: Towards a politics of conviviality', *Science as Culture*, **15** (2), 123–38.

Københavns Kommune (2015a), *3. Bynatur i København 2015–2025 (2015–0219237)*, accessed 21 March 2019 at https://www.kk.dk/indhold/teknik-og-miljoudvalgets-modemateriale/02112015/edoc-agenda/e0a9935f-3899-4462-917c-38c002f61dc9/2bbb8421-4961-4dc3-be82-0a68905b57ab.

Københavns Kommune (2015b), *Bynatur i København. Strategi 2015–2025*, May, accessed 11 November 2016 at http://kk.sites.itera.dk/apps/kk_pub2/pdf/1447_EDTTg7TXgO.pdf.

Latour, B. (2005), *Reassembling the Social*, Oxford: Oxford University Press.

Lave, J. (2011), *Apprenticeship in Critical Ethnographic Practice*, Chicago: University of Chicago Press.

Law, J. (2017), 'STS as method', in U. Felt, R. Fouché, C.A. Miller and L. Smith-Doerr (eds), *The Handbook of Science and Technology Studies. Fourth Edition*, Cambridge and London: MIT Press, pp. 31–57.

Loughran, K. (2016), 'Imbricated spaces: The high line, urban parks, and the cultural meaning of city and nature', *Sociological Theory*, **34** (4), 311–34.

Mangset, M. and K. Asdal (2019), 'Bureaucratic power in note-writing: Authoritative expertise within the state', *The British Journal of Sociology*, **70** (2), 569–88.

Schechner, R. (1968), '6 axioms for environmental theatre', *The Drama Review: TDR*, **12** (3), 41–64.

Schön, D.A. (1984), *The Reflective Practitioner: How Professionals Think in Action*, New York: Basic Books.

Suchman, L.A. (1987), *Plans and Situated Actions: The Problem of Human–Machine Communication*, Cambridge: Cambridge University Press.

Swyngedouw, E. (2006), 'Circulations and metabolisms: (Hybrid) natures and (Cyborg) cities', *Science as Culture*, **15** (2), 105–21.

Tavory, I. and S. Timmermans (2014), *Abductive Analysis*, Chicago and London: University of Chicago Press.

Timmermans, S. and I. Tavory (2012), 'Theory construction in qualitative research: From grounded theory to abductive analysis', *Sociological Theory*, **30** (3), 167–86.

Vinck, D. (2012), 'Accessing material culture by following intermediary objects', in L. Naidoo (ed.), *An Ethnography of Global Landscapes and Corridors*, Rijeka, Croatia: InTech, pp. 89–108.

Zuiderent-Jerak, T. (2007), 'Preventing implementation: Exploring interventions with standardization in healthcare', *Science as Culture*, **16** (3), 311–29.

Zuiderent-Jerak, T. (2015), *Situated Intervention: Sociological Experiment in Health Care*, Cambridge, MA: The MIT Press.

14. Staging urban design through experimentation

Birgitte Hoffmann and Peter Munthe-Kaas

INTRODUCTION

In 2009, the global population in urban areas exceeded the population in rural areas (United Nations (UN) 2010); more than one-third of the European population now lives in urban areas (EU Commission 2011). Around the world, and especially in the EU, the city is seen as vital to sustainable development, social cohesion, and economic development.

> As the global population becomes increasingly urbanized, cities have emerged as the dominant arenas to address the grand challenges facing humanity. Problems associated with climate change, economic under-development and social inequality are essentially urban in character. And so are their solutions ... The burgeoning realization that 'business as usual' will no longer do has prompted a search for alternative ways to organize, plan, manage, and live in cities. Experimentation promises a way to do this, gaining traction in cities all over the world as a mode of governance to stimulate alternatives and steer change. (Evans et al. 2016: 1)

However, traditional systems of urban governance and planning are struggling to handle the challenges, where uncertainties and controversies are multiplying (Callon et al. 2009).

Urban planners are 'embedded within current institutions and structures that have coevolved with earlier, typically modern practices, and are yet surrounded by the pressures of new, complex, contemporary problems that require novel practices' (Lissandrello and Grin 2011: 226). Although new, deliberative forms of planning such as neighbourhood regeneration programmes are emerging, urban governance is becoming increasingly fragmented, and market oriented procedures and new networks of interest across the local, national and international levels are gaining influence on urban designs and development. Hence, the planners' position as experts of urban development is being challenged by the focus on citizens and their experiences of urban life. And while scholars such as Healey (1998) argue that this development creates new platforms for democratic and active participation, critical voices such as Flyvbjerg (1998), point out the risk of decisions being made in new networks outside the planners' professional expertise and citizens' democratic influence.

Both perspectives put pressure on planners, who are increasingly expected to serve as active mediators to transform and develop urban areas. Planners can no longer legitimately claim to simply use professional practices to translate political information into new directions for urban spaces. This sub-politization of society underlines how politics is more widely distributed in the planning system (De Vries 2007). It also means that it is no longer possible to operate with clear notions of the 'good city' that can be attained through spatial planning alone. The current definition of 'good' is much more processual and fluid, with signifiers such as 'liveability' and 'sustainability' that are far from clear. Thus, urban planners today find themselves in a situation where their professional expertise is no longer sufficient to develop the city, and it is difficult to separate planning from politics and power (Graham and Marvin 2001; Lissandrello and Grin 2011).

In such dynamic and fragmented environments, it becomes crucial to develop new professional and democratic planning practices that can operate within the complexities, controversies and uncertainties present in late modern societies. Whereas controversies were previously seen as threats to be reduced and/or removed, they (at least in theory) are increasingly seen as starting points for exploring how society can develop, and as tools for enriching the emerging world (Callon et al. 2009). Hence, planners need to find new points of departure in urban development by opening up existing social-technical assemblages and experimenting with alternative urban futures.

In this chapter, we explore staging in this context of planning for sustainable urban futures. To discuss staging in relation to the field of urban planning, we focus on the intentional setting of the stage as part of a democratic governance process. We explore the 'crafting of the invitation' by asking how 'stakeholderness' is choreographed in the staging and how the invitation opens up sociotechnical urban spaces to become 'disputed' to engage a multiplicity of actors in experimenting with the creation of new urban futures.

Our discussion of staging in this chapter is rooted in the field of public governance, where democratic participation is axiomatic and where municipalities have different forms of obligations to govern for sustainability. The contribution of this chapter is largely to reveal the democratic implications of different staging strategies. However, our critical discussion and analysis of the invitation to staging has relevance beyond urban planning and governance, particularly in the fields of open innovation and user-centred design, which are predicated upon participatory ideals, as well as in design efforts focused on sustainability. Hence, we claim that in all staging processes, the question of equipping stakeholderness should be pivotal.

In the next section, we develop our approach to staging. Based on a pragmatic understanding of the city as an assemblage, we draw on new agonistic perspectives of democracy to reveal implications for planners in their staging of urban development projects. Applying a Thinging lens, we develop our analytical frame focusing on how the crafting of the invitation to the staging matters for stakeholderness: Who is invited on the stage and what futures can be explored? Afterwards, we present three pioneering cases that show ways of working with staging in urban planning. In two rounds of analysis, we explore invitations to the staging process, including which staging devices are used to equip stakeholderness and to open up opportunities to experiment with future urban living. We conclude on the learnings in relation to staging in urban planning and point to perspectives for the development of staging as an approach in design and innovation.

SUSTAINABLE URBAN DEVELOPMENT AS A THINGING PRACTICE

The staging approach in this book refers to the activities of intentionally and strategically drawing together a sociotechnical assembly and comprises a repertoire of ways to configure and facilitate interactions across actors and objects while taking into consideration the uncertain, temporal, and agential aspects of design and innovation (see Chapter 2, this volume).

To discuss this definition in relation to the field of urban futures, we draw on a pragmatic tradition whereby places or situations are made political when they become 'disputed' or open to discussion—and thus to controversy:

> The radical departure pragmatism is proposing is that 'political' is not an adjective that defines a profession, a sphere, an activity, a calling, a site, or a procedure, but it is what qualifies as a type of situation. (Latour 2007: 815)

Ideas from pragmatism have had an impact on what Patsy Healey (2009) called relational approaches to planning. These ideas link to the increased interest in complexity and the 'wicked problems' that planning is currently expected to address, and point to new understandings of what planning can become.

> This understanding of practice suggests a distinctively counterhegemonic or democratizing role for planning and administrative actors: the exposure of issues that political–economic structures otherwise would bury from public view, the opening and raising of questions that otherwise would be kept out of public discussion, the nurturance of hope rather than the perpetuation of a modern cynicism under conditions of great complexity and interdependency. (Forester 1993: 6, paraphrased in Healey 2009: 284)

Opening the city up to alternative interpretations can be described as a movement that in Bruno Latour's words, reflects a transition whereby 'matters of fact' which are understood as truths begin to transform into 'matters of concern' which are linked to the many different notions that can exist about any urban area (Latour and Weibel 2005). This pragmatic and sociotechnical approach to urban development points to an important shift in the planner's role. Reproducing fixed notions of development based on expert knowledge is no longer sufficient when we want to challenge what is taken for granted and to explore alternate possible futures. Planners need to be able to handle complex sociotechnical assemblages with multiple interests in play.

This line of thinking is mirrored in agonistic approaches to democracy and governance. Here, consensus is understood as a temporary result of the existing hegemony, a stabilization of power that always involves some form of exclusion (Mouffe 2012). An agonistic perspective recognizes the controversies in society and the exclusions that must unavoidably occur in any decision, rather than trying to hide them behind notions of rationality. Neutrality is not an option when adopting an agonistic perspective; rather, the role of planning is to open up processes to actors and perspectives formerly rendered invisible:

> For 'agonistic pluralism', the prime task of democratic politics is not to eliminate passions from the sphere of the public in order to render a rational consensus possible, but to mobilize those passions towards democratic designs. (Mouffe 2012: 38).

From this perspective, democracy is a process in which facts, notions and practices in society are constantly challenged and explored and where places are created where confrontations are possible and where a diversity of actors can participate. Thus, the planner's task becomes one of constructing stages where different urban understandings can be debated and new realities can be explored (Björgvinsson et al. 2012; Pløger 2004).

This leads to our main concept in our approach to staging. The idea of Things draws on the Nordic word for physical meeting places for disputes and political decision-making (Latour 2004). We use this concept to emphasize how urban places and processes can be worked with as spaces of controversy and democratic exchange, rather than problems that have correct (engineered) solutions. This involves challenging hegemonic notions or 'matters of fact' and inviting multiple actors into Thinging practice (Ehn 2008; Binder et al. 2011)—a process of making the ordinary controversial to imagining, conceptualizing and creating the new.

By framing urban design as a Thinging practice, we find that new spaces can be opened for the political to be (re)introduced in urban planning, and we explore how staging can develop as an approach to create Things, which can engage a multiplicity of actors in the development of the cities of the future.

Hence, in our analysis we are interested in exploring how the setting of the stage engages a multiplicity of actors and gives space to uncertainties and controversies to enable experimentation with alternative urban futures.

CRAFTING THE INVITATION: ONTOLOGICAL CHOREOGRAPHY AND STAKEHOLDERNESS

In current planning, the 'affected stakeholder' holds a prominent position regarding the purpose and potential outcomes of urban planning processes. In dominant practices, participatory processes are staged as 'hearings', with 'citizens' meetings' as the main devices which in principle are open to everyone, yet typically involving only a small group of citizens, either those with very specific stakes or 'the usual suspects'. This practice not only positions actors in a reactive role, but also excludes multiple actors and perspectives.

Inspired by Metzger (2013), we question who is to be considered a stakeholder when staging processes, and highlight that this is not only an epistemological challenge, but also a fundamentally ontological issue:

> Thus, diverging apprehensions of who are ascribed the property of being 'affected' may radically shift the composition of the group of stakeholders who should be considered to legitimately hold a right to partake in deliberations within the planning process (e.g. are the citizens of the Maldives to be considered as legitimate stakeholders in Vancouver's planning process, and thus to be given a right to voice in that process, seeing that they are 'affected' by the potential sea-level rise resulting from global warming which may be the result of a development of road traffic infrastructure in any urban area such as Vancouver?). (Metzger 2013: 783)

Hence, planners do not just convene stakeholders, they generate and foster stakeholderness by the way projects are framed and organized. Thompson (2005) describes these processes of manipulating interests and attachments through reality-crafting practices as 'ontological choreography'. In other words, to stage anything in the urban context, the planner (more or less consciously) performs acts of ontological choreography.

In their work with democratic design experiments, Binder et al. (2015: 162) emphasize invitations as central to framing a proposal for the new, while also opening the process for co-creation:

> Crafting an invitation to participate in a democratic design experiment is an active and delicate matter of proposing alternative possibilities just clearly enough to intrigue and prompt curiosity, and, on the other hand, to leave enough ambiguity and open-endedness to prompt the participants' desire to influence the particular articulation of the issue.

Based on this line of argument, we find that there is a critical need in planning to explicitly work with invitations into staged processes. This emphasizes the need to analyse the crafting of the invitation to the process of experimenting with futures; choreographing which actors (and actants), forms of knowledge and realities are included in the staging; and, subsequently, what the staging excludes.

STAGING IN URBAN PLANNING: THREE CASES

To explore and discuss this focus on staging, we introduce three innovative cases of planning projects from Copenhagen that in different ways staged new urban futures. The first case involves a large-scale urban development project explicitly staging a sustainable urban future laboratory, making it important as a global inspiration. The two other cases are interesting as they involve staging experiments with the future sociotechnical urban fabric in the form of prototyping respectively long-term intervention processes.

We selected these projects from our portfolio as action researchers who partner with practitioners to develop new urban approaches. We describe and analyse the cases based on our own research, primary literature such as plans and websites, and research performed by other scholars. Presenting three cases in a single chapter naturally requires simplification, and the descriptions here are very brief; nevertheless, we use the diversity to explore how staging in municipal planning unfolds on different scales, over different timeframes, and in different ways. Each case points to central aspects of staging related to the outlining of new planning practices based on creating Things with a special focus on the crafting of invitations.

Based on the frame developed in the previous section, we analyse how (if) setting the stage creates space for uncertainties and controversies in each case, thereby enabling the creation of Things that can engage a multiplicity of actors around matters of concern to produce alternate urban realities. Our analyses focus on the crafting of invitations to the staging process and the application of staging devices. In short, we ask how stakeholderness is choreographed and how invitations open up opportunities to experiment with possible new urban futures.

Setting the Stage: Choreographing Stakeholderness

In this section, we introduce the cases and analyse how stakeholderness is choreographed through the crafting of invitations and application of staging devices.

Nordhavn

The redesign of Nordhavn (www.nordhavnen.dk), a former industrial harbour area in Copenhagen, is a high-profile development project to realize the Danish capital's branding as a liveable and sustainable metropolis.

In 2009, By & Havn, a developer owned by the municipality and the state, crafted an invitation to participate in an open international competition to contribute to a master plan for 'the sustainable city of the future' to be developed over a 50-year period.

The Nordhavn development is a huge and complex project, which of course cannot be reduced to a simple process of inclusion and exclusion. As Blok states, 'Zoning laws, land prices, construction materials, energy technologies, risk analyses, building standards, stylistic fashions, user habits, and so on—all of this (and more) is brought together, worked upon, modelled and modified in and beyond the architectural office' (Blok 2013: 13). However, with an architectural competition as the main staging device, architects and engineers were equipped to lay out the frames for future development in the master plan. Subsequent competitions have since been held to translate the plan into concrete designs of buildings and infrastructures for the initial phase of the project. This traditional urban development practice involved inviting professional architects with their knowledge and imaginaries of urban sustainability, urban living, developers' rationales, etc. to participate. Their designs framed the later invitation to developers to buy plots and hire additional architects and engineers and entrepreneurs to detail and build the local designs. Lastly, potential users were invited to realize the final designs in practice by moving in. In this way, the staging of this high-profile urban development gave agency to certain experts in urban design and sustainability. Citizens and other user groups as well as other forms of knowledge and perspectives that could contribute to exploring alternative ideas of the sustainable city of the future, including what it could mean to 'live sustainable futures', were not equipped with stakeholderness in the design process.

Herman Bangs Plads

In 2012 in the municipality of Copenhagen, planners from the technical department responsible for waste and recycling partnered with planners from one of the decentralized urban regeneration programmes and developed an idea to combine the desire to place a waste collection station in a local square with underfinanced citizens' wish for a recreational space in a rather dense urban area. This case is an example of the municipality staging a participatory design process on a local scale by using prototypes as an approach to experiment with urban design.

The Herman Bangs Plads project was aimed at developing and testing a new kind of 'waste collection point' in an otherwise 'empty' square to increase recycling. At the same time, the local regeneration programme wanted to transform the square into an attractive urban space. Combining these interests, the planners crafted the invitation with a temporary 1:1 prototype as the main device to stage a possible future for the square and test the potential of a multifunctional collection point.

A small design consultancy was hired to build a cheap model for the three-month long experiment. The prototype itself consisted of a shipping container converted into a swap shop, a pallet staircase to a small terrace on top, a small playground next to the container and some additional containers for recyclables. The prototype was not just physical, however. The regeneration programme initiated some events at the location about up-cycling, as well as social and cultural events.

The prototype served as an invitation to local residents to explore and test the facility by fostering interaction with the facilities and the other actors. The planners learned that the prototype also acted as a device to stage constructive informal dialogues with citizens beyond the official citizens' meeting format, and facilitate learning across municipal sectors. The idea as well as the prototype itself were products of expert design, but the experimental approach opened up opportunities for locals to engage and explore.

Tåsinge Plads

The overall aim of a regeneration programme in Outer Østerbro was to support placemaking by engaging local citizens and businesses in improving recreational functionality and creating a new local identity for a rather featureless urban area at the fringes of the city. This case is an example of how a local green area was staged by diverse temporary interventions to explore opportunities for future redesign and use.

The area became one of the first Danish climate adaptation cases to relocate storm-water handling from below ground pipe systems to nature-based green infrastructures integrated in the urban landscape.

The invitation to storm-water to co-produce the future use of the area added resources to realize a local greening strategy as well a new identity as a 'climate neighbourhood'. At the same time, by imbuing nature-based water infrastructure (and water planners) with stakeholderness, the future of the area became framed around a nature-based design and giving space to storm-water in case of heavy rain events.

One main space was Tåsinge Plads, originally a small green area mainly used by dog walkers that rested on top of old concrete wartime shelters surrounded by asphalt streets and parking lots. The planners invited locals to explore the area's future layout and use by devising a series of interventions, and, over the course of two years, different actors became engaged in performing temporary activities at the space.

Facilitated by the urban regeneration programme, planners, university students, nature enthusiasts and local citizens created temporary and open-ended art installations and workshops to perform alternative futures of the area formerly known as 'dog shit hill'. For example, different pieces of urban furniture were built to enable residents to experience potential uses, and the 'instant hygge (cosiness) initiative' brought people together for Saturday breakfasts and social events. Likewise, flea markets, urban gardening and other initiatives staged new possibilities and awareness of the space.

The different activities served as invitations to different actors and forms of knowledge. The events and installations engaged many residents, including young families who seldom attend traditional citizens' meetings. The process opened up various ways for locals to participate, including making it legitimate to 'just' be a spectator, hence inviting diverse groups of citizens to experience local urban life in new ways.

Summing up

In this section, we have briefly introduced three different cases of staging in the urban setting, illustrating how invitations to the staging of urban futures actively choreograph actors (and actants) in different ways, using devices to give 'voice' to certain actors and exclude others. Importantly, those who perform staging in urban planning need to give deliberate attention to how staging equips certain actors with stakeholderness.

The critical quest of sustainability gives rise to the specific task of giving voice to the 'future generations' that became framed as central actors in the UN's sustainable development concept titled 'Our common future' (World Commission on Environment and Development 1987). Taken seriously, this task is rather interesting and should pique curiosity and experimentation. This leads to the connected inquiry of how to give agency to sustainable urban futures from an existing urban reality.

Setting the Stage: Opening up New Urban Futures

In this second round of analysis, we examine how setting the stage gives space to uncertainties and controversies, thereby enabling the creation of Things that can engage a multiplicity of actors around the development of new urban futures.

Urban planning is inherently about the ongoing effort to constructively shape a desirable urban future, and a growing number of cities all over the world actively position themselves as front-runner innovators in sustainability (Bulkeley and Broto 2013). However, critical questions are raised as to whether these planned urban futures will adequately address the global crisis. Furthermore, based on the above, we advocate for the need to challenge the existing hegemony and develop more agonistic approaches to explore different visions of sustainable futures.

This line of thinking can also be seen in the works of Jean Hillier (2008), who argues that cities should work more with emergent developments in society, thus thinking less in terms of what we already know, but more in terms of what we do not yet know. The perspective is a multiplanar practice of spatial planning, working with 'a broad trajectory of possible scenarios, developed and debated democratically, inclusively and deliberatively, to "rehearse" possible futures and their perceived advantages and disadvantages to actants' (Hillier 2008: 29–30). Hence, what we find interesting is when staging opens up opportunities to explore futures not yet known.

The sociomaterial urban fabric and its many everyday practices is simultaneously stabilized through network effects that are reproduced every day, and destabilized by minor controversies that emerge among actors or by larger overflows (Farias 2010). In this way, we see all urban planning and design as redesign. This involves navigating between existing realities and unknown futures to empower transitions of existing non-sustainable practices. Hence, planners working to stage alternative futures must also address the existence of pathways in the form of existing sociomaterial urban layouts and practices, and take up the challenge of deliberately working with destabilizing matters of fact.

Furthermore, because sustainable urban futures are not yet known, it is a specific challenge to create and qualify Things where multiple actors can engage in controversies and experiments with alternative future realities. Thus, rehearsals of possible futures become a key element in staging sustainable development. This directs attention to discussions about how uncertainties and alternatives are addressed to open up opportunities to explore sustainable futures. In our case analyses, we focus on how experimentation through concepts such as laboratories, prototyping and temporary interventions are applied as staging devices to empower designs with the capacity to open up opportunities to explore possible futures.

Nordhavn

The design specifications positioned Nordhavn 'as an urban "laboratory" for testing various "cutting-edge" green technologies, implying that experiences gained from this locality will be readily transferable to other contexts'; hence, the international competition performed Nordhavn as a case of 'context-free design thinking' (Blok 2013: 12).

The main invitation to stage the 'sustainable city of the future' focused on professional designs of urban development that were both innovative, attractive and marketable. The competition framed the possible visions by focusing on the existing strategy of Copenhagen with a strong focus on technological green innovations (Blok 2013), thereby focusing the enquiry into sustainable living as a process of greening infrastructures, for example, by saving ground water by using local salty ground water to flush toilets (Hoffmann 2018).

As Blok (2013) noted, such plans are never entirely solidified and may be re-opened for public-political scrutiny once architectural designs begin to be inscribed. However, the Nordhavn development left little space for uncertainty and (economic) risks, particularly because the financing of the new Copenhagen Metro was dependent on the sale of the publicly owned harbour plots. By now, the first phase of Nordhavn has been built and the residents have moved into an urban area with a design rather similar to those of other development sites.

On one point, the Nordhavn development inscribes a rather alternative urban future, outlining the '5-minute city' that highlights cycling, walking and public transport as an easy choice. Unfortunately, the development does not address the controversies of car practices and parking spaces. Even though the development of Nordhavn underlines sustainability as a matter of urban concern, the design of the first phases were framed more as a 'test facility' to mainstream a sample of green technologies than as an experiment with future sustainable living. In short, the Nordhavn project so far is mainly based on the same urban hegemony as most other developments of this type in Copenhagen and other cities.

Herman Bangs Plads

At Herman Bangs Plads, urban planning professionals initiated a dialogue with local actors through the device of a physical prototype in an urban space. This enabled multiple actors to engage with the future of the square and opened up public deliberation that in many ways exceeded what traditional plans and processes could have performed. Still, the staging was framed quite strongly by the idea of combining local recycling with urban life. In other words, the Thing that was created brought the city in to discuss how to shape the future of the square, but the 'what' had already been largely decided. In this way, we see parallels to user-driven innovation processes, where users are consulted on the design and details of a product (Sanders and Stappers 2008). However, the prototype also opened up local interaction in more serendipitous ways. For example, the swap shop and playground space became meeting places that fostered interaction between the richer middle-class citizens living in villas on one side of the street, and the people living in high rises deemed 'ghettos' by the Danish government on the other side.

Tåsinge Plads

The temporary interventions at Tåsinge Plads qualified the debate of the future use of the site by proposing a series of different possible designs and activities.

The interventions—picnic breakfasts, urban gardening, building of furniture and art performances—were rather open-ended to embrace possible interpretation by the participants and to spur further development. For example, the initial art event placing hundreds of large sticks attracted substantial attention to an area that otherwise seemed overlooked by everyone but the dog walkers. The sticks, not proposing any specific use, stimulated bypassers to be aware of and interested in exploring the area, thereby provoking a reframing of 'dog shit hill'. In this way, the staging of interventions created a Thing that opened up uncertainties about the place and surrounding urban life and built new relations around the reshaping of the local area. The planners, however, black boxed one high-profile controversy from the very beginning. They guaranteed not to reduce the overall number of parking places, as they feared that any attempt to open up the existing car practice to debate would stall a constructive process.

SETTING THE STAGE: THINGING FOR SUSTAINABILITY

Our analysis points to the fact that staging is taking place in all three cases. The high-profile Nordhavn urban development project was staged as the 'sustainable city of the future', the Herman Bangs Plads project was staged as an innovative combination of waste collection and recreational space and the Tåsinge Plads project was staged as a space-making process to redesign the local urban space and urban life. The cases clearly reveal the invitation as a critical element of staging in urban Thinging processes and yield some important insights into the crafting of invitations.

An architectural competition was the main staging device in Nordhavn, which involved deploying a well-known practice whereby experts produce plans and realities. This could have surfaced a diverse set of ideas that could have enabled Nordhavn to be developed as a global statement of sustainability as a matter of urban concern. However, the green technology focus framed the experts' imaginations, and the traditional practice of using a closed forum of experts and representatives to choose a winner narrowed the process. The requirements of establishing a successful urban development with secure revenue from the sale of high-value plots seemingly made it undesirable to open up the project to too much uncertainty. The urgent need to address sustainability causes us to question how the long-term development period of 50 years for the project can be used to simultaneously pursue short- and long-term goals by implementing safer, readily acceptable ideas during the initial phases, while staging experiments to rehearse radical other visions of sustainable urban futures that can inspire and qualify later phases.

In the Herman Bangs Plads project, the use of the 1:1 prototype invited local citizens to experiment with new urban infrastructures and urban life. Some planners as well as local citizens were sceptical of using scarce public space for waste collection. Yet, the temporality of prototyping over a three-month period enabled people to engage in new experiences that led individuals to reconsider their hitherto understandings of both waste collection and recreational space. Moreover, even though the planners framed the experiment, the test design was flexible enough to include local interpretations. After the experimentation phase, a group of local citizens participated in the design of the concrete liveability-recycling station model for the square, which also was to become a design for the city in general.

In earlier work, we described this kind of staging as a strategic design experiment, serving the purpose of drawing attention to and sparking deliberation about a professionally defined agenda (Munthe-Kaas and Hoffmann 2017). We see this as a possibility to open existing urban planning practices to enable both citizens and professionals to imagine more radically different futures.

In the Tåsinge Plads project, we want to highlight both the open-ended interventions and the long-term period for rehearsing possible futures. The different sets of events invited a diverse group of local residents to explore what could be done in this urban space and what future urban life could be like. The planners played a traditional role by initiating the process, but used the open-endedness as well as the timescale to invite others to participate in the staging. Hence, the layout was opened up for reinterpretation along with the stakeholderness of citizens in planning and performing ownership of public areas. In the end, a statutory tender to rebuild the space was issued. Although many inputs produced during the staging process were included, parts of the local engagement were lost during formal approval and development processes, which tend to be long and 'invisible' to citizens. The planners explained that this specific experience has informed a new planning practice of integrating processual tender requirement about participation.

Perhaps unsurprisingly, our analysis highlights the difficulties of staging for alternative futures in the large-scale and high-profile case of Nordhavn. The two smaller scale projects with shorter timeframes were easier to manage and integrate into existing planning practices. Moreover, less was at risk, even though local residents and existing sociomaterial practices still had to be addressed in the processes of staging new urban futures. This raises a challenge to be dealt with, since large urban development projects are extremely important to support sustainable transitions. Based on the arguments made in this chapter, our main advice to planners is to work strategically to stage pilot projects as ways to support Thinging, as pilot projects can be used to prompt concrete experiments to expand realities and invite additional actors to commit. Along these lines, in large, long-term projects such as Nordhavn the use of phases should be deliberately used to issue invitations for experimental pilot projects. Planning for sustainable urban futures must be seen as an iterative process of strategically staging Things to change urban realities, whereby planners play an important role in crafting invitations while considering how staging frames the stakeholderness of present and future perspectives.

CONCLUSION

In this chapter, we have explored staging in the urban planning context. As defined by Pedersen et al. in this volume (Chapter 2), we view staging as activities of intentionally drawing a sociotechnical assembly, a repertoire of ways to configure and facilitate interactions across actors and objects while taking into consideration the uncertain, temporal and agential aspects of design and innovation. From this perspective, staging is an integrated element of urban planning practices, and we have shown how the concrete choices made by planners in the staging process play a central role in shaping urban futures.

We have constructed a theoretical framing with 'Thinging' as our main approach to underline staging as a political process. Using the concepts of 'ontological choreography' and 'agonistic pluralism', we offer a framework to critically and constructively discuss the setting of the stage. Hence, based on our development of a Thinging Practice we emphasize the need to carefully work with the 'invitation' to the stage to equip stakeholderness and to open up the process for uncertainties and controversies to constructively play a part in designing.

Our analysis of innovative cases of planning for urban futures confirms the need to develop planning and urban design processes that are able to deal with the huge challenges confronted by global society. In a world where the development of sustainable urban practices is of vital importance, we find that these considerations around how to choreograph staging are central to this development. The field of urban planning is dominated by hegemonic modernist ideas that frame how lives can be lived. Opening up planning through staging processes that attempt to make the ordinary controversial, particularly through physical experiments involving multiple stakeholders, can point to alternative ways of living in cities in the future.

Our brief analysis of three cases reveals how different staging approaches involving the use of different devices led to specific inclusions/exclusions, and resulted in different levels of success in terms of opening up opportunities for alternate sustainable futures. By highlighting the role of invitations, we have revealed the embedded stakeholderness of the concrete cases of staging. We argue that planners can use a Thinging approach as a way to direct staging efforts towards the conscious inclusion and equipping of actors and agency to create openings for new ideas regarding future sustainable practices.

A Thinging perspective addresses the democratic potential in staging. It implies explicitly crafting invitations to the staging process to open up opportunities to experiment with urban futures by choreographing which actors will participate and how, and deciding which realities will be included.

This critical discussion of staging is relevant for design processes in general. The invitation choreographs: Who gets to be on stage and who does not? What can be dealt with and what cannot? How are existing realities destabilized to open up alternatives? And how is the rehearsing of futures qualified? To create Things that open up black boxes and pave the way for inquiries into future solutions and realities, designers need to explicitly and carefully craft invitations to the staging of design processes.

REFERENCES

Binder, T., E. Brandt, P. Ehn and J. Halse (2015), 'Democratic design experiments: Between parliament and laboratory', *CoDesign*, **11** (3–4), 152–65.

Binder, T., E. Brandt, J. Halse, M. Foverskov, S. Olander and S.L. Yndigegn (2011), 'Living the (codesign) lab', paper presented at *Nordes 2011*, Helsinki.

Björgvinsson, E., P. Ehn and P. Hillgren (2012), 'Agonistic participatory design: Working with marginalised social movements', *CoDesign,* **8** (2–3), 127–44.

Blok, A. (2013), 'Urban green assemblages', *Science & Technology Studies*, **1**, 5–24.

Bulkeley, H. and V.C. Broto (2013), 'Government by experiment? Global cities and the governing of climate change', Transactions of the Institute of British Geographers, **38** (3), 361–75.

Callon, M., P. Lascoumes and Y. Barthe (2009), *Acting in an uncertain world: an essay on technical democracy*, Cambridge, MA: MIT Press.

De Vries, G. (2007), 'What is political in sub-politics? How Aristotle might help STS', *Social Studies of Science*, **37** (5), 781–809.

Ehn, P. (2008), 'Participation in design things', in *PDC '08: Proceedings of the Tenth Anniversary Conference on Participatory Design*, Bloomington, IN: ACM, pp. 92–101.

EU Commission (2011), *Cities of Tomorrow—Challenges, Visions, Ways Forward*, Luxembourg: European Commission—Directorate General for Regional Policy.

Evans, J., A. Karvonen and R. Raven (2016), *The Experimental City*, Abingdon: Routledge.

Farias, I. (2010), 'Introduction: Decentring the object of urban studies', in I. Farías and T.B. Bender (eds), *Urban Assemblages How Actor–Network Theory Changes Urban Studies*, Abingdon: Routledge, pp. 1–24.

Flyvbjerg, B. (1998), *Rationality and Power—Democracy in Practice,* Chicago: University of Chicago Press.

Graham, S. and S. Marvin (2001), *Splintering Urbanism: Networked Infrastructures, Technological Mobilities and the Urban*, London and New York: Routledge.

Healey, P. (1998), 'Building institutional capacity through collaborative approaches to urban planning', *Environment and Planning A*, **30**, 1531–46.

Healey, P. (2009), 'The pragmatic tradition in planning thought', *Journal of Planning Education and Research*, **28** (3), 277–92.

Hillier, J. (2008), 'Plan(e) speaking: A multiplanar theory of spatial planning', *Planning Theory,* 7 (1), 24–50.

Hoffmann, B. (2018), *Salt Flush – User Perspectives on Secondary Water Quality and Sustainability* (in Danish), Copenhagen: Aalborg University.

Latour, B. (2004), 'Why has critique run out of steam? From matters of fact to matters of concern', *Critical Inquiry*, **30** (Winter), 225–48.
Latour, B. (2007), 'Turning around politics: A note on Gerard de Vries' paper', *Social Studies of Science*, **37** (5), 811–20.
Latour, B. and P. Weibel (ed.) (2005), *Making Things Public: Atmospheres of Democracy*, Cambridge, MA: MIT Press.
Lissandrello, E. and J. Grin (2011), 'Reflexive planning as design and work: Lessons from the Port of Amsterdam', *Planning Theory and Practice*, **12** (2), 223–48.
Metzger, J. (2013), 'Placing the stakes: The enactment of territorial stakeholders in planning processes', *Environment and Planning A* **45** (4), 781–96.
Mouffe, C. (2012), 'An agonistic model of democracy', in M. Jahn (ed.), *Pro+agonist: The Art of Opposition,* Minneapolis: Northern Lights.
Munthe-Kaas, P. and B. Hoffmann (2017), 'Democratic design experiments in urban planning—Compositionist design in practice', *Codesign* **13** (4), 287–301.
Pløger, J. (2004), 'Strife: Urban planning and agonism', *Planning Theory*, 3 (1), 71–92.
Sanders, E. and P.J. Stappers (2008), 'Co-creation and the new landscapes of Design', *Codesign* **4** (1), 5–18.
Thompson, C. (2005), *Making Parents: The Ontological Choreography of Reproductive Technology*, Cambridge, MA: MIT Press.
UN (2010), *World Urbanization Prospects: The 2009 Revision*, New York: United Nations, Department of Economic and Social Affairs.
World Commission on Environment and Development (1987), *Our Common Future*, Oxford: Oxford University Press.

PART VI

Reflections – how staging is understood and used

15. Taking the metaphor of theatre seriously: from staging a performance toward staging design and innovation

Dominique Vinck and Mylène Tanferri

INTRODUCTION

In this book, *Staging Collaborative Design and Innovation*, the concept of staging is proposed as the core concept, taking some of its roots in the theatrical experience. In Chapter 1 of this volume, we identified some useful characteristics of the staging process to study design and innovation. We insisted on its help to avoid reducing design activities to a socio-cognitive process involving experts acting according to a pre-established method. On the contrary, the notion of staging orients our attention towards the casting process, stage framing and the resources it offers to participants. Thus grounding on the theatrical metaphor opens new perspectives, enriching our understanding of design and innovation (D&I). In this chapter, we dive into this metaphor in order, first, to point to its risks and limitations and then to enrich the notion thanks to both a field study on staging a performance and a literature survey on theatrical practices. In conclusion, the chapter broadens the perspectives explored in this book.

THE RISK WITH METAPHORS

In this book, staging is proposed as an analytical and operational concept. Analytically, metaphors make it possible to organize a network of analogies to explore the investigation data, and reveal new avenues of analysis. However, using metaphors is always risky. If metaphors allow us to apprehend reality and create new perceptions of it, they may also become the focus of attention at the expense of the investigated phenomenon. Metaphors have a cognitive value by permitting analogical reasoning. Our conceptual system is metaphorical in nature (Lakoff and Johnson 1980). Metaphors sometimes make it easier to understand or even discover aspects that we would not have thought of, like staging as a metaphor to focus the attention on the preparation of D&I. But they risk missing the phenomenon altogether if they give the illusion of a thoughtful description, providing explanation and better understanding, especially if it is brilliant. They could also lead to categorical confusion (Ricoeur 1993) by replacing both the phenomenon and a more literal approach and vocabulary. Using metaphorical language to describe the phenomenon under study in place of a more literal vocabulary, rather than providing clarity, may sometimes hinder it. Worse, metaphors can become a protective shell that allows us to go through the studied phenomenon without being affected. Then, they become a cliché that stops the investigation instead of pushing it further. They may also attribute qualities that come from the metaphor to the studied phenomenon, obscuring it by introducing things that are not there.

All of this argues for using theatrical metaphors with caution, without losing sight of the reality they are supposed to represent. This suggests investigating both the phenomenon that we want to shed light on and the phenomenon stylized by the metaphor. It is then a question of making a detour to other research fields, not to establish a point-by-point comparison, but to open up new heuristic leads on the preparation of D&I activities. Thus, in place of using a metaphor to summarize and replace these processes, we propose to explore it and ground it in order to enrich our research tools.

With this perspective, the chapter proposes two contributions: first, to look at the actual preparation of a performance in order to enrich our understanding of staging. Going through the specifics of a particular performance will allow us to question the metaphor, to give it an explicit basis, so it can retain an analytical force. Second, the chapter proposes to look at the history of theatre to highlight the diversity of its practices and thus avoid being trapped by a simplistic metaphorical reference. In reality, theatre corresponds to a great diversity of situations that "the metaphor of the theatre" (in the singular) is likely to conceal.

LEARNING FROM THE OBSERVATION OF STAGING A PERFORMANCE

During 2016–2019, we had the opportunity to follow and observe the preparation of the Fête des Vignerons, a special celebration organized in a small city in Switzerland to celebrate the life and work in the local vineyards and thank their workers. The Fête has been celebrated for more than two centuries and is organized every 20–25 years. In 2019, its design and preparation took seven years, leading to a major performance involving 5,500 voluntary players on stage.[1] The region's inhabitants are invited to get involved as performers in the show and the troupes, choirs and brass bands learn the texts, music and dances of the show. The Fête thus involves a whole region and participants from various backgrounds coming together to produce the show.

Staging as an Emerging Figure

While the show respects a tradition of celebrating the life and work of the vine, it is, in fact, re-created in each generation. The Brotherhood calls on artists to create a text, music and a show. Even if many elements of the narrative structure are repeated from one edition to the next (e.g. the division into seasons, Grecian–Latin deities), each generation introduces new elements linked to what has happened in between (e.g. a political revolution, a war, the destruction of vineyards by new parasites, the arrival of agricultural machinery and the protection of biodiversity issues).

Poets, composers, costume designers and decorators, as well as choreographers, rebuild this tradition in each generation. Originally, the poet and his text formed the basis on which the other participants would work. For a long time, staging came last. If we follow details of the Fête organization from different editions, we observe that the figure of the stage director emerges at the beginning of the twentieth century. This figure becomes increasingly important. The director orders and organizes the show, and decides on costumes, props, sets and choreography with the corresponding creators. During this process, propositions are made, specifics of each speciality are presented and their potentials and limitations are discussed to be progressively included in the director's creative plan. Following these processes and negotiations from an historical perspective is, in and by itself, an interesting opportunity to observe the prominence of the stage director in the 2019 edition.

As we observe a growing importance of staging D&I, and, after a self-generated dynamics or leadership of a domain-specific expert, does a new leading professional figure (a stage director for participatory design) emerge?

The Progressive Shaping of Artistic Decisions

In 2009, the Brotherhood set up a preparatory commission to look into the artistic aspects of the show. This commission made a critical assessment of the previous celebrations and outlined ideas for the next one. The commission realized that in 20 years the world has changed so much (climate change, the spread of digital technology, new managerial practices, new public expectations) that practically everything had to be reinvented. The Brotherhood defined a general direction for the celebration, and consulted with different generations of the cultural scene. In 2011, the Brotherhood announced its wish for a modern, popular and dreamlike show, combining emotion, tradition and anchorage in the professional, economic and cultural reality of the moment, while remaining a great fresco that speaks of the vine and life, of human work and the earth, and of how to pass on a clean land to future generations.

The Brotherhood then began looking for a designer and artistic and stage director. In 2013, it chose Daniele Finzi Pasca, an internationally renowned director for his mastery of large-scale events (notably the closing ceremony of the Turin Olympics in 2006), who is known to produce emotionally engaging shows. Once recruited, the stage director immersed himself in the daily life of the vineyard and its local traditions. From this experience, he then proposed a narrative frame inspired by the gestures of the winegrowers while shaking up certain traditional elements he considered too sexist, militaristic or linked to foreign deities.

Designing the stage and performance thus involves different stakeholders (the Brotherhood, the professionals of the artistic world, but also people talking in name of the tradition) and the director, who brings a specific experience of show production which he hybridizes with an impregnation of local life and tradition. Prior to staging, the Brotherhood and the director recuit a creative team, according to skills, styles and experience: two poets and two composers from the region, plus the director's previous collaborators (a composer, a costume designer, a scenographer, a visual content designer and a choreographer). The director also uses his previous approach to work collaboratively (without anyone – poets, musicians or others – taking precedence), towards the collective design of images and atmospheres from which everyone then conceives bits of text, costumes, music, stage elements or choreography. Through brainstorming, the team produces a lot of material, which is then examined, criticized and reappropriated by the other creators. Along the way, the results become difficult to attribute to anyone. Poets, costume designer and musicians are inspired by the ideas and comments of others, while the scenographer draws a stage from an idea of the atmosphere. Here we can see that the director first stages the creative process, which then results in a storyboard and the very staging of the show.

In this process, the director must deal with a series of unexpected events, among others the replacement of the scenographer. He also has to deal with the Brotherhood, particularly in terms of budget, agreement on the elements to be retained, abandoned or transformed from the tradition, the recruitment and involvement of volunteer performers, and the inclusion of television crews for a live broadcast of the show and the future production of a souvenir film.

For staging D&I, we note: even before the start of the stage design, the director organizes the process relying on his resources, network and methods; he orchestrates and negotiates with heterogeneous stakeholders precisely how staging will ensue; casting and staging needs to be done for the design activity, which is then confronted with various elements that it needs to include.

Staging the Scenery and the Performance

The show actually emerges from the staging process also by taking into account various possiblites and limitations. It is not limited to the casting of the performers and a few arrangements on a pre-existing stage. In fact, what renders the Fête particularly interesting to study is that its stage is ephemeral: it does not pre-exist and must be reinvented for each edition even if it always takes place on the market square. Its integration in the city evolves greatly from one century to the other from an open parade to a closed stage with access tickets. In a critical evaluation of the 1999 celebration, the 2019 director ruled out the idea of opening the stage to the lake as it was done previously to avoid elements of the surrounding scenery (mountain views, lake and boats) competing with the show. He wants to plunge the audience into a story to be lived together. The arena should cut the audience off from the surrounding world. The director and the scenographer propose to build a nest in which the audience will be enveloped, a space where the story of the winegrowers can be told in "private" – but for 20,000 spectators! An elliptical arena is chosen to create a show both grandiose and intimate, able to host a crowd while maintaining a feeling of closeness to the show.

Goffman (1974), working with stage, theatre and drama as conceptual helping hands, developed the notion of framing to get insights about the interrelations between a framed activity and its surroundings. Framing is a way for participants to define the situation by answering the question "what is it that's going on here?" (1974: 8). The notion offers two interesting elements. First, a frame – definition of what is going on – must be maintained and its changes carefully planned. It cannot stand by itself, requires ongoing care and needs to be modified smoothly to avoid open disagreement. Second, it points to the fact that every activity needs to be demarcated, somehow, from the ongoing flow of other activities. This demarcation can be only rhetoric, but framing can take more substantial forms. For example, entering the arena is already a way to perceive a demarcation not only from the surrounding animations but also from other environmental events such as the sun's reflection on the lake water, which would interfere with the daytime show. The arena then works as a framing device which at the same time continues to be interrelated to its outside. The start of the play also has to be demarcated by three strikes of the gong and turning up the music. But even if it is to put the outside aside for a moment, what happens in the surroundings nonetheless continues to play its role and can even threaten to disrupt the framed activity. This theatrical metaphor allows us to follow staging as way to work toward the definition and boundaries of the show.

The audience's attention is a resource that the director integrates within a series of divergent imperatives: immensity and intimacy, massive – and therefore potentially slow – movements of performers vs. a rhythmic show. To organize spectators' attention, instead of designing a twofold space with the stage on one side and spectators on the other, the stage director and the scenographer draw a central stage and four adjacent stages, halfway up the arena, connected by a corridor and four majestic staircases. The audience, wherever they sit, should be overwhelmed by actions close to them, before their eyes, but also behind their backs and at their sides, and thus be immersed in the show. As the frame of the arena is perceived as unsufficient to fully capture the spectators' attention, the show should provide other frames and potential involvments. Spectators should be able "to touch the show with their fingers", says the director. Staging therefore also includes the shaping of specific places for the spectators, while stage, backstage and stands must provide the technical support necessary to create effects of surprise and wonder. Monumental theatrical objects or hundreds of performers need to appear or disappear smoothly thanks to entrances and exits, trap doors and opening staircases to avoid breaking the tempo and thus spectators' involvement in the fiction. It's all about capturing viewers' attention, "who may otherwise pause to look at their smartphones and digital social networks". Specific elements are recruited to design boundaries, to capture attention, and to create a separation from the ordinarity of everyday life compared with the magic of the show.

For D&I, we underline: staging is designing the stage, not only its content, spaces and places for further users; staging is seting up compromises between contradictory requirements, among which is the audience's attention; participants work toward the framing of the activity and setting its surroundings, which are both interdependant and demarcated; stage, backstage and users' places must provide the support necessary to capture users' attention and to produce the expected results.

Putting a Stage into Existence

Boundaries or frames can be thin and volatile like sound, speech or technical ways to mesmerize the audience or solid and enduring like the arena. The arena construction is negotiated with more entities as it encounters not only the concrete marketplace and its specificities, but also the inhabitants, city council and other actors, who have a say about its building in public space. Moreover, even if a play is to be staged inside the arena, it continues to happen somewhere, using elements that are enrolled in the show and will endure after it.

Indeed, from the dream of a giant nest to its construction, the road is long and winding.[2] Adding to the creative team, other actors progressively appear: architects, civil engineers, insurance companies, construction companies, suppliers of stage machinery, sound equipment, etc., whose ideas, know-how, resources and requirements influence the design, but also associations (environmentalists, shopkeepers and residents) and public authorities (civil protection, bearers of society's requirements in terms of safety, sustainability and accessibility) who all have a say in its final design. Between them circulate countless digital or printed drawings. These intermediary objects support the confrontations and coordination between actors. The director works to reach compromises between aesthetics, technical and budgetary feasibility, and scenographic relevance. Technical equipment itself is negotiated and reconfigured. Two years were needed to design, validate (of performance, feasibility, costs) and construct the stage. In the meantime, technologies evolved, which condition the possibilities and constraints for artistic creation and the ability to stage 6,000 people simultaneously.

Thus, design and construction faces material constraints, among which are the marketplace properties: a 17,500 m^2 space in the city centre, surrounded by buildings, sloping towards the lake, and whose basement is a spider's web of water, sewage, gas, electricity and telecommunications networks. Since the building must be solidly anchored in the ground, this subsoil influences the design of the show; since the location of the 200 piles is dictated by the free zones of the underground network, the location of the show's sound masts and accordingly the sound design of the show will be affected.

Scenographic thinking is complex. To give rhythm to the show, the director, the scenographer and the choreographer define the appearance and disappearance of various elements and numerous performers from the main stage to backstage and design the arena accordingly. The spacing also depends on aesthetic issues and is reflected in, among others, the trapezoidal form of the side stages, which in turn conditions the lighting and sound system. Lighting professionals and acousticians are thus enlisted in the design of both the building and the show. Articulating physical constraints and scenic choices, the scenographer draws horizontal stages and stands, despite a 3 per cent incline, involving a stage floor at 3 metres above ground level. He takes this opportunity to include an access ramp for the performers under the central stage. Accessibility requirements for impaired persons also influences design, while spacing anticipates potential events, such as evacuation in case of storm or terrorist attacks (exit number and size), or the planned dismanteling once the celebration is over (rented elements and reusable materials). Thus, instead of having a temporal dissociation between spacing, show design and staging, as well as the involvement of the corresponding professions, one observes an interweaving of activities that condition each other.

For D&I, we note: staging involves many concrete elements with which to negotiate, including many emerging actors, issues and changes; spacing, design and staging interweave and condition each other.

Staging the performance with performers and props

Staging not only includes the coordination of various actors around the arena building, it also encompasses crafting a whole series of details, in and around the show, taking into account available resources and multiple constraints. It implies planning for the audience in different ways (e.g. when defining acceptable sound levels for various ages or with the seamless inclusion of costumed security staff to avoid disturbing the dreamlike atmosphere of the show) and includes the involvement of props, costumes, decor and traps, backdrop, set pieces, machinery, lighting, and the video animation of the 800 m^2 of LED floor on the main stage and their respective technicans. Staging also depends on the content of the show whose tradition is reinvented. With the Brotherhood, the director defines a list of must-haves (attachment to the land, love of the nation, traditional songs or characters), and rethinks them according to today's society. Public expectations then constitute resources and constraints, but staging has to consider not only the opinions but also the skills of non-professional performers and singers.

Composers incorporate the abilities of local choristers in the very content of the music score. Traditionally, songs and music composed for the Fête have nourished the regional choral repertoire, so expectations are already high. Staging the show nourishes the tradition by enriching its musical heritage as much as the choral tradition nourishes and supports the show. The compositions will not only be used for the show but also for the next 20 years to revitalize the choral milieu.

Two years before the show, the director recruits the performers. Apart from a few professionals for specific roles, the majority (5,500) are volunteers. The recruitment is open to anyone from the region, regardless of age and skills. Choristers are recruited from existing choirs, while a casting is organized for the dancers (1,800). The average age is 45, which is high for performers who have to parade, dance or sing, and endure 20 days of performance. Accepting everyone whatever their aptitude, the choreographer, assisted by professional dancers and gymnasts, auditions people in groups of 100 performing a small choreography, which is filmed and then evaluated to assign volunteers to a scene adapted to their abilities. Staging also implies redesigning certain scenes according to the number of people able to follow the rhythm, and be flexible and graceful.

With D&I too, staging both depends on local resources and enriches them, no matter whether it is users' expectations or the know-how of the inventors of the solutions.

Rehearsals as the Central Process of Staging

Staging does not stop at the choice of the stage set, performers and costumes, it also goes through numerous rehearsals. As performers and singers are not professionals, staging implies important logistics and planning for more than a year of rehearsals for 5,500 people. Of course, logistics constraints weigh on staging and the final performance: rehearsal cut-off times for children's choirs; provision of costumes and props; equipping everyone with headsets and 300 singers with high-frequency microphones; managing performers' availability, etc. Rehearsals are a major staging component: they allow for all perfomers to prepare and make sure they will be able to perform as expected. Performers, singers, as well as stage managers, technicians and operators, incorporate, integrate and modify along the way what will have to take place at the final performance. For more than a year, the director and an increasing number of assistants take care of the rehearsals.

In addition, through rehearsals, the show is re-adjusted and re-oriented. During rehearsals of the choirs, brass bands, gymnastics groups and dance groups, the creative team observe the performers' actual skills and involvement and the real performance of the technical equipment. Rehearsals involve scenes being shortened, musical compositions being cut, even if already learned by the choirs and dancers, and changes in choreography and backstage movements, as well as daily improvements of the sound system tested in real conditions of use.

Observing staging work thus offers the possibility of emphasizing the importance of rehearsals, preparing people and action, including training, coaching, briefing and debriefing. Staging is also a long process of rehearsal, re-design and adjustment, not only the preparation up to the first public performance but continuing from show to show. After each performance, collective debriefing sessions go back over the show and define the changes to be made. More than 20 rehearsals for each scene, plus five dress rehearsals, stabilized the show before its performance in front of an audience, but, in fact, the show is never the same from one time to the next. Rehearsals, then, are a collective learning process where certain points of stabilization emerge. Around these "fixed" points, which are the work of the stage manager and topper to keep in place (not the director any longer), performers know they can adjust, adapt or improvise: small incidents, fatigue, enthusiasm, may lead to a clandestine touch-up on the costume to adapt it to the requirements of the performance, discreet mutual aid between performers during the show may happen, as small improvisations or improvements as a claim to their role as actors.[3]

The final learning from this observation is that: staging goes through numerous rehearsals – this would be valid for D&I too; staging consists in preparing props, action, but also performers; it implies (backstage) activities taking place between rehearsals and between performances (see Chapter 6, this volume).

WHAT WE LEARN FROM THE THEATRICAL EXPERIENCE

The Fête is perhaps not representative of theatre history. In this section, we will report on some of the theatre's practices and evolutions around the material space of the stage in order to enrich the metaphor and its use for staging D&I. We must be careful when speaking of theatre because it is not a homogeneous reality, any more than D&I.

Stage and Scenography as Frames for the Performance

Since antiquity, the theatre space has been a configured space which interplays with staging activites, performance im/possibilities and audience involvement. Open sky or not, it isolates the performance from noise, world events and competing interactions. It is also structured internally by differentiating the stage on which the action takes place (involving decor, flattage and traps, wing curtains shaping the perception by the public) from the room where the spectators come to watch the performance, and the backstage area where the performers and the elements are prepared. The theatrical space is thus configured in a specific way, but with countless variations (raised or lowered stage, surrounded by the hall or in a face-to-face encounter with the audience) which affect staging, performance and audience involvement.

However, not all theatres can be reduced to this type of enduring sociomaterial arrangement. Street theatre for instance does not benefit from such isolation, which then needs the actors to shape an attentional space, interrupting the courses of action of passers-by to transform them into an audience. In the theatre of Bauhaus, spaces allowed the action to surround the spectators or moves through space, leading to new audience–performer relationships (Kaprow 1960). In environmental theatre (Schechner 1968), there is no segregation between the stage and the hall, between performers and audience. In the 1960s, the arrangement evolved (Surgers 2005), by taking the show out and mixing it with urban space, allowing or forcing the public to move. At some point, we may say that any space (including the web) becomes a stage as soon as one person acts while another observes (Brook 1977).

In contemporary theatre, a variety of scenographic solutions are explored to blur the notion of a singular stage space by connecting distant spaces (e.g. on-stage audio-visual projections captured outside the theatre). This mixture of technologies mediates the here and now of the show with elsewhere and from another time. Directors also reinvent the spectators' space by equipping them with video-headsets, and immersing them in the action. Using screens on the stage enlarges the space by introducing (video)images, sometimes fictitious such as Pepper's Ghosts (optical illusion producing a transparent image, without apparent support). The image creates real and imaginary spaces, makes the spectator travel (another country, the inside of a human body, the thoughts of a character, etc.), plays on scales (e.g. a close-up of a face makes the spectators feel small (Picon-Vallin 1998), temporalities and realities (Pluta 2011). By superimposing themselves on the stage, these images produce a sense of presence or immersion in another reality. Staging transforms the spaces that the spectators are confronted with, for instance relying on the solitude of the spectator to facilitate the fright, through personal interfaces, or, on the contrary, on being used by many people simultaneously who meet in a virtual theatre (e.g. a play on second life) and experience a shared adventure. It draws the audience into a hybrid space that seems present and sensitive (Dixon 2007).

The concept of staging underlines that the (D&I) stage results from a creative work involving closure and internal spacing, connecting spaces, shaping of attention space and any sociomaterial arrangements framing user–performer relationships and innovative performance.

Actors at the Heart of the Theatre, but Non-humans Too

The performers' presence on stage is at the core of the theatre. Thus, staging implies their casting, the allocation of roles, as for staging D&I, but also the interpretation of a text (in D&I maybe specifications), a proposal or starting instructions – like in improvisational theatre. The interpretation depends on the collective, its accumulated experience and dynamics. Performers' enrolment varies from staging to staging. When plays are performed by a troupe used to working together and with their director, collective know-how is a resource on which to draw. If the collective does not pre-exist, their interactions begin during rehearsals. As performance depends on the interaction among them, their casting, the team building and the working conditions influence staging, creativity and the performance. Everyone lends both their own perspectives and materiality to the staging and performance and is transformed by them when they leave.

Introducing new "actors" and their technicians

Besides the relations between performers, staging also deals with non-human entities and associated technicians, their role in staging and their interactions with performers. Often hidden, the machinery is either at the service of the scenic illusion or exhibited (e.g. as a sign of progress). From the sixteenth century onwards, the stage was structured to allow for change and surprise (appearance and disappearance).[4] Staging was the site of scientific events that arouse curiosity and celebrate power:[5] smoke screens, fireworks, fountain or mirrors games, watchmaking systems and digital technologies. Scholars had then to present their knowledge according to the theatrical taste of courtiers (Tkaczyk 2017), creators and audiences. Part of the play is delegated to scenographic elements and their experts (researchers, stagehands, engineers, technicians and programmers). Directors mobilize new materials and devices on stage: phono-dynamic costumes, masks equipped with lamps instead of eyes and megaphones instead of mouths (Berghaus 2002), screens that intensify the political message. With the 1960s American avant-garde, hybrid forms appeared, mixing performance, cinema, dance and plastic arts (Picon-Valin 1998) to produce expanded theatre and filmic happenings. Thus, staging involves photography, film capture and projection, radio and audio recordings, data capture (on the bodies of performers), digital processing, telematics and new restitution interfaces.[6] These elements serve as either background for performers or as performing elements. The technicians contribute to staging beyond the provision and refinement of the equipment. Thus the "performing group" (Schechner 1968) is expanded. During the performance, performers and technicians support themselves mutually and react to each other. "Technicians should be a creative part of the performance. It is even possible that elements will be made and rehearsed separately and that the performance itself is the arena where competing elements meet for the first time" (Schechner 1968: 59). The same applies for D&I staging.

Staging then relies on the machinery and ingenuity of the stagehands. These scenographic elements and their technicians are not simple instruments that carry out the director's orders; they actively participate in the interpretation of the drama and its transformation (Boenisch 2006). They are active devices, constitutive of the creative process and the performance, that produce, in turn, artistic content. Their potentialities and constraints are partly unknown to the director. Surrounded by a diversity of specialists, staging consists of a dialogue with professions in order to get a coherent result. Since the 1960s, creations have become hybrid and the distinction between contributions complicated. These collaborations presuppose the invention of intermediary objects capable of encouraging the expression and conflation of points of view and supporting their agreement: models, prototypes, simulation and digital file exchanges are becoming common practice.

Tinkering with new materials and technologies

Technology needs to be articulated with the performers. While it can help to make stage dreams come true, directors question its modes of presence and coexistence. Staging is then a collective exploration of possibilities, diversions and tinkering to make it their own. Directors confront the materials and techniques they try to master, but which also impose unexpected constraints, failures or unexpected effects. They try to disconnect their ego and submit to the materials and what they communicate (e.g. the inflatables whose way of moving is linked to the chaos of the air). Materials force people to understand their behaviour before they can lead them; they suggest material answers to artistic questions. Staging is also hacking the technology and using it in a way that was not planned.

However, inventing and experimenting new techniques is limited because performance requires reliable technology. But staging is an opportunity to test alternative uses of technologies already validated, and to gain insights through making and experiencing. Staging is then exaltated by what emerges from materialities put to the test and their dialogue. From their manipulation and articulation emerge new artistic realities that surprise their creators, while digital equipment blurs boundaries between components and enables new assemblages. Staging, far from being the translation of a text into materialities and persons, is a collective exploration from which the text and computer code emerge. The theatre is a stage of intermediality (Chappe and Kattenbelt 2006) that lends itself to multiple browsing. The scene has become a composed reality that questions the performer's presence (Auslander 1999) and representation, and the performance's non-reproducibility (Benford & Giannachi 2011). Digital technologies multiply, transfigure, replace and increase the elements captured from the scene (e.g. image, sound and movement of the performers)[7] and lead to a performance which is also an effect of calculation processes (Causey 2016). Staging confronts performers with their double filmed and projected on screen; they must interact with their image on screen, a photograph that stops time or a film that accelerates or slows down time, compresses events, which questions the temporality of the action on stage. Performances on stage could also be transferred to a non-human double (e.g. an animal) on screen or on stage (a remote-controlled object or a humanoid robot). Staging plays with realities, which occurs within D&I.

With Chatterbot[8] performances, liveness seems real due to the engagement of a conversation with these entities, considered able to perform live. Stage environment is thus manipulated by artist-programmers. Bringing to life multiple materialities challenges staging, which manages degrees of presence or liveness on stage. Staging then confronts practical insights in the ontology of performance (Dixon 2007). It "augments" reality by adding new dimensions and makes visible the fiction present in the sociomaterial arrangements, in the use of techniques and in the staging of life. It shows that fiction is part of reality and can transform it.

Staging then experiences a technological increase with interfaces to interact with. Actors become cyborgs themselves, equipped with sensors and sensory feedback. The idea of a self confined to a skin envelope loses its meaning

(Paterson 2009). Sound mix and amplification erase distinctions between performers, or even confuse them when it is transferred to an avatar. Staging creates a presence that depends on sociomaterial assemblages, generating a feeling, somatic responses and kinaesthetic empathy (Reason and Reynolds 2012). The introduction of humanoid robots on stage also creates a disturbing strangeness. Staging then requires writing a text for the robots, taking into account their personality.

The Spect-actors

This overview of theatrical explorations populates the world of staging with performers, technicians, materials and technologies. However, one component still needs our full attention: the spectators. Looking at it, we can push the accepted boundaries between who makes and performs, or benefits from and attends the performance to explore further what agency and participation means. The distribution of roles and power plays among other things in the attention and immersion of the spectators in the show. Traditional theatre puts them in a passive attitude of visual and auditory reception. They attend a show presented frontally and from which they are physically separated. But staging already included them through the shaping of their perception (see Chapter 12, this volume, by Jensen talking about silenced actors). Spectators have never been reduced to the role of an eyewitness. They are summoned to watch, to listen, to feel multisensory proxemics experiences, or having personal thoughts but in a framed way. An expressive/silent audience could support the performers depending on the play. Creating harmonious reactions is sought after, but traditional theatre barely explores potential interactions unlike the fairground theatre.

The distribution of roles and power also concerns the interactions that make up the show. In classical theatre, the play takes place between the performers on stage. The audience is sometimes left aside, sometimes evoked or challenged en masse, but without being given the power to act. Normally, the spectators respect the rules of behaviour: arriving on time, not leaving their seats until the show is over, displaying (dis)approval within well-regulated patterns of applause, silence, laughter, tears and standing ovations.

With contemporary developments, immersion was explored through staging different events happening at the same time, attracting and diverting spectators' attention to different spaces behind, above and below. In the Fête, we have seen that each event competes for the audience's attention while no spectator can see everything. The spectators realize they will inevitably miss parts of the show and that others may have a different experience. Performers and technicians, used to project their play and voice to an entire audience, can be disturbed by this reframing. They can express a lack of control because every spectator would assemble a different spectacle – in D&I, a different user experience.

Contemporary reflections about spectators' roles and places have explored the distribution of agency and constitution of boundaries between performers and spectators. Intermedial happenings and the environmental theatre since its political and radical form of the 1960s display a variety of experimentations with public involvement, including performer–spectator interactions. These experimentations also worked to create disruptions to awaken spectators from the dream of fictional theatre, making acting and staging visible as such. They sought to inactivate usual identification and immersion effects and give the spectators their critical thinking abilities back (Brecht and Bentley 1961). Theatre of engagement has tried to create an active audience from the group of people assembled but not prepared to act as a group (e.g. by stimulating them to question themselves and to interact with other spectators). Since the 1970s, post-dramatic theatre has been pushing interactional dynamics further by hybridizing performance and writing in such a way that staging becomes collective stage writing. All this questions the users' participation in D&I (see Chapter 7, this volume, on transition design by Clausen and Gunn, and Chapter 6, this volume, on action-nets by Dorland). In the environmental theatre, initially for reasons of political commitment, and later for popular entertainment in a consumer society (Nelson 1989), staging implies a flexible approach to the performer–audience relationship (Schechner 1968) and an alternative to public withdrawal and voyeurism. The notion of environmental theatre emphasizes the constitution of an environment for interaction, involving both performers and audience and becoming a self-generating dynamic (see Chapter 13, this volume, by Johansen and Lindegaard on situated interventions).

The deployment of video and communication technologies on stage opens up an important development in this respect (Laurel 1993). From the 1990s onwards, Internet and mobile media made it possible to involve dispersed audiences as spect-actors. Inspired by video games, staging in the game theatre confronts them with enigmas, where they have to make choices and collaborate. They may give the performers instructions for their performance or, equipped with joysticks or tablets, they can animate their audio-visual avatar on stage and act on its life. The director acts as a moderator. The scene is not written in advance, which makes the performance unpredictable. It becomes then a collective, constantly evolving form of writing, in which no one in particular is in charge. The roles are less differentiated. Public participation implies a relative loss of control on the part of the director and raises the problem of the aesthetic coherence of the performance. Spect-actors can form a group generating its own dynamics, which becomes co-author of the artwork that emerges from the interactions.

These experiences raise the question of public engagement, which is sensory and emotional but also a social, political, cultural involvement, linked to the realities of society (Lavender 2016). Engagement in the show (as user) or in creating and staging it in a fluid way, i.e without forcing or pointing fingers, asks what is needed to make sure people are willing to share and participate in sensemaking. Providing options to express opinions about an already framed experience is not the same as becoming so involved that we feel responsible for the collective dynamic. Staging may assign new roles to spect-actors and allow them to experiment with new ways of forming an audience (Dewey 1927) (e.g. political interventions, smart mobs, collective dances or audio walks, by coordinating the emergence of both performers and audience) in spaces where ephemeral stages can emerge, enhanced by the collective performance that reappropriates the public space in a different way. Staging creates a relational network between performers and (dispersed) spect-actors (Bourriaud 1998) that forms an actor-network or action-nets (see Chapter 4, this volume), an emergent agency that these practices perform (Binder et al. 2015).

For the Fête des Vignerons, staging included voluntary performers; part of the spectators were their family and contacts and the staged event was another way to relate and interact. Moreover, as volunteers, they worked at the intersection between public and performers. Selected amongst the local population, they engaged in rehearsals, the fitting of their costumes, learned the proposed choregraphy, and contributed to designing a multifaceted un-singularized event, multiplicating voices and perspectives about the show not only through their performance, adaptations and interpretations but also through conversations and media postings. In this sense, they also defined what the Fête is.

FURTHER POTENTIAL OF THE THEATRICAL METAPHOR FOR DESIGN

Theatrical metaphor was proposed to think about design activity (Brandt and Grunnet 2000) and to reinvigorate the democratic impulse of collaborative and participatory design, replacing the text and its interpretation by the staging of sociomaterial conditions for exploratory and controversial issues and Thinging as sociomaterial assemblies that evolve over time (Telier 2011), facilitating public participation, contested expressions of matters of concern, contradiction, oppositions and issue formation. The preoccupation is also to expand the range of participants (outside pre-specified ones) for issues and sociomaterial assembling. This follows the programmatic call for rethinking design (Latour 2008), as "drawing things together" (Latour and Weibel 2005) to render design more compositionist (Latour 2010).

While the use of a metaphor can be risky if it simplifies, idealizes and homogenizes the studied phenomena, close examination of theatrical practices, showing their diversity and evolution, reinforces the relevance of the intuitions that lead to its use. This confirms the shift of attention from text and method to the preparation of a sociomaterial assembly (casting of human and non-human actors) that should favour the deployment of an interactional activity and the emergence of collectives and solutions to meet challenges. Charlotte Louise Jensen (Chapter 12, this volume) points to the who and what is staging spaces and the role of the "humble stage director" who wants and needs to learn from the performers. She also argues that material aspects act on the practices at least as much as the meanings and competences. Birgitte Hoffmann and Peter Munthe-Kaas (Chapter 14, this volume) shed light on the invitation to the stage. They show how staging creates stakeholderness. Looking at the crafting of the invitation to stakeholders, the stakeholderness and the ontological choreography stakeholders would follow, they propose to make visible the democratic implications of staging strategies. Their questioning would also include staging devices to be used for equipping stakeholderness. Thus, it's relevant to look, like Signe Pedersen and Søsser Brodersen (Chapter 5, this volume), into the "backstage" work and the iterative nature of staging, facilitating negotiations and reframing. Jens Dorland (Chapter 6, this volume) considers how staging structures the dynamics allowing the creation of processes and empowerment (action-nets); staging is then described as unfolding impacting events. The improvisation theatre or game theatre metaphor is even more relevant for thinking about D&I where the performance differs for each enactment. The performance is not fully defined and mastered (with timecode and rehearsing for embodying routines), especially if the boundaries of the stage and the focus of attention are challenged as shown by Ask Greve Johansen and Hanne Lindegaard (Chapter 13, this volume). This questions the attribution of cues to performers, to the equipment and set elements, and to spect-actors or users.

While the metaphor is fruitful and well exploited, several aspects of theatrical practices may be underestimated. As we underlined through the Fête des Vignerons case study and the literature review on theatre transformations, two elements stand out: (1) the key role played by rehearsal (the period during which the components of the performance are explored, transformed and assembled, including the human actors who will perform on stage or in the audience); (2) the multiple forms of audience engagement. On these two issues, the metaphor of theatrical staging still has a true potential that deserves to be exploited. In that sense, Dorland (Chapter 6, this volume) opens up promising prospects. In his analysis, he adds to backstage strategic planning and frontstage plays activities occurring between events, such as the production of a document or artefacts to be enacted during the events. But this may be explored further with the role of rehearsals: rehearsing is also an activity occurring between planning and events and between events. It includes roles, performers, technical operators, but also the play itself (a kind of action-net) which is enacted later. "[S]taging is an act of trying to configure an action-net in a specific way", writes Dorland. There are activities to prepare people, to give them the ability to play their role, to empower them. But rehearsals go further than motivating participants; this concerns also the production and negotiation of a collective sensemaking practice and the progressive enmeshment between a creative vision and its actual forms through interplays and interpretations. Here the connection with public engagement needs to be explored further. Staging is a process which creates a space for action (an action-net allowing new action). It requires also the production and circulation of knowledge across sociotechnical spaces and mutual learning (see Christian Clausen and Wendy Gunn in Chapter 7, this volume) in order to get mastery and to improve and stabilize a collective performance.

In this sense, the metaphor may lead to a confusion between staging the D&I process (in theatre: orchestrating the creative process) with staging the product until it enters daily practices (in theatre: staging the show). The theatrical metaphor may be used more profitably to analyse the staging of D&I (the orchestration of the creative process) instead of analysing the staging of the final performance (the development of a product until its routine use). Staging may seem to concern mainly the show, but with the Fête des Vignerons we observed two staging processes: one for the creative process and one for the performance. This difference is relevant for D&I, as staging occurs not only once a product is released and during its design, but also upwind during the staging of D&I. Exploring actual theatrical practices instead of taking the metaphor at face value offers fresh perspectives on possible ways to study and rethink the design activity and to reinvigorate its democratic impulse as "drawing things together". Thinking about staging opens up new debates about public engagement, expressions of matters of concern, and negotiation to form relevant issues, publics and sociomaterial assembling by pointing to the various stages where participation may occur.

NOTES

1. See an excerpt on: https://www.fetedesvignerons.ch/media/les-extraits-du-spectacle-2019/ (accessed 13 May 2020).
2. The stage as projected in 2017 (https://www.fetedesvignerons.ch/it/video/larene-de-la-fete-des-vignerons-2019/), during its building (https://www.facebook.com/radiochablais/videos/540498679786060/?v=540498679786060) and before the performance (https://www.facebook.com/watch/?v=923613724636401) (all urls accessed 13 May 2020). The design was "slightly" modified during the staging.
3. Regarding rehearsal see: https://www.fetedesvignerons.ch/media/les-choeurs-de-la-fete-des-vignerons-2019/, https://www.fetedesvignerons.ch/media/les-repetitions-du-spectacle/ (accessed 13 May 2020).
4. See also the art of trickery and effects with the boulevard theatre and fairground shows.

5. Galileo was the organizer of festivities for the Medici dynasty; the Academy of Sciences in Saint Petersburg was in charge of court shows; geodesy, hydraulics and mechanics were summoned at the Palais de Versailles for its gardens and festivities, and the spectacular affirmation of royal power. The avant-garde theatre of the 1920s shows the "new man" and art nouveau in a context of industrial and societal change.
6. For example, video-headsets, surround sound, headphones, Bodycoders, smartphones, exoskeletons, robots, implants.
7. Software for real time management, sound processing and spatialization, lights, interfaces between machines.
8. "A specific, conversationally interactive form of the numerous 'bots' (short for 'robot') that reside in cyberspace, from search engine agents to 'warbots, channelbots, spambots, cancelbots, clonebots, collidebots, floodbots, gamebots, barbots, eggdrop bots, and modbots'" (Dixon 2007: 491).

REFERENCES

Auslander, P. (1999), *Liveness: Performance in a Mediatized Culture*, London: Routledge.

Benford, S. and G. Giannachi (2011), *Performing Mixed Reality*, Cambridge, MA: MIT Press.

Berghaus, G. (2002), 'The futurist body on stage', in N. Roelens and W. Strauven (eds), *Homo orthopedicus. Le corps et ses prothèses à l'époque (post)moderniste*, Paris: L'Harmattan.

Binder, T., E. Brand, P. Ehn and J. Halse (2015), 'Democratic design experiments: Between parliament and laboratory', *CoDesign*, **11** (3–4), 152–65, accessed 13 May 2020 at https://doi.org/10/f3pqj2.

Boenisch, P. M. (2006), 'Aesthetic art to aisthetic act: Theatre, media, intermedial performance', in F. Chapple and C. Kattenbelt (eds), *Intermediality in Theatre and Performance*, Amsterdam/New York: Rodopi.

Bourriaud, N. (1998), *Relational Aesthetics*, Dijon: Les Presses du réel.

Brandt, E. and C. Grunnet (2000), 'Evoking the future: Drama and props in user centered design', in *Proceedings of Participatory Design Conference (PDC 2000)*, Palo Alto, CA: CPSR, pp. 11–20.

Brecht, B. and E. Bentley (1961), 'On Chinese acting', *The Tulane Drama Review*, **6** (1), 130–36.

Brook, P. (1977), *L'Espace vide. Écrits sur le théâtre*, Paris: Éditions du Seuil.

Causey, M. (2016), 'Postdigital performance', *Theatre Journal*, **68** (3), 427–41.

Chappe, F. and C. Kattenbelt (eds) (2006), *Intermediality in Theatre and Performance*, Amsterdam: Brill Rodopi.

Dewey, J. (1927), *The Public and its Problems: An Essay in Political Inquiry*, University Park, PA: Pennsylvania State University Press.

Dixon, S. (2007), *Digital Performance: A History of New Media in Theater, Dance, Performance Art, and Installation*, Cambridge, MA: The MIT Press.

Goffman, E. (1974), *Frame Analysis: An Essay on the Organization of Experience*, Boston, MA: Northeastern University Press.

Kaprow, A. (1960), *Assemblages, Environments, and Happenings*, New York: Harry N. Abrams.

Lakoff, G. and M. Johnson (1980), *Metaphors We Live By*, Chicago: The University of Chicago Press.
Latour, B. (2008), 'A cautious Prometheus? A few steps toward a philosophy of design (with special attention to Peter Sloterdijk)', Keynote lecture for the Networks of Design. Falmouth, Cornwall, UK.
Latour, B. (2010), 'An attempt at a "Compositionist Manifesto"', *Review of New Literary History*, **41**, 471–90.
Latour, B. and P. Weibel (2005), *Making Things Public: Atmospheres of Democracy,* Cambridge, MA: MIT Press.
Laurel, B. (1993), *Computers as Theatre*, Reading, Boston, MA: Addison-Wesley Publishing Company.
Lavender, A. (2016), *Performance in the Twenty-first Century: Theatres of Engagement*, London/New York: Routledge.
Nelson, S. (1989), 'Redecorating the fourth wall: Environmental theatre today', *The Drama Review*, **33** (3), 72–94, accessed 13 May 2020 at https://doi.org/10/d4txsb.
Paterson, M. (2009), 'Haptic geographies: Ethnography, haptic knowledges and sensuous dispositions', *Progress in Human Geography*, **33** (6), 766–88.
Picon-Vallin, B. (ed.) (1998), *Les écrans sur la scène,* Lausanne: L'Âge d'Homme.
Pluta, I. (2011), *L'Acteur et l'intermédialité. Les nouveaux enjeux pour l'interprète et la scène à l'ère technologique*, Lausanne: L'Âge d'Homme.
Reason, M. and D. Reynolds (eds) (2012), *Kinaesthetic Empathy in Creative and Cultural Practices*, Bristol: Intellect Ltd.
Ricoeur, P. (1993), *The Rule of Metaphor*, Toronto: University of Toronto.
Schechner, R. (1968), '6 axioms for environmental theatre', *The Drama Review*, **12** (3), 41–64, accessed 13 May 2020 at https://doi.org/10/ffv2p9.
Surgers, A. (2005), *Scénographies du théâtre occidental*, Paris: Armand Colin.
Telier, A. (T. Binder, G. De Michelis, P. Ehn, G. Jacucci, P. Linde and I. Wagner) (2011), *Design Things*, Cambridge, MA: The MIT Press.
Tkaczyk, V. (2017), 'Which cannot be sufficiently described by my pen: The codification of knowledge in theatre engineering 1480, 1680', in M. Valleriani (ed.), *The Structure of Practical Knowledge*, Berlin: Springer, pp. 77–115.

16. Navigating with people and objects – strategic concerns

Yutaka Yoshinaka and Christian Clausen

INTRODUCTION

In the shaping of design and innovation, heterogeneous concerns are brought into play in the course of collective interaction and transformation. How such purposeful and collective, heterogeneous shaping is enabled through actionable strategies in the staging of design and innovation is the central topic of this chapter. Here, we highlight and elaborate on this notion through our reading across chapter contributions in this book. The chapters demonstrate designing as distributed enactments, through the circulation and unfolding of ideas, objects and actors in the process. Rather than the actions of a sole designer, designing is hereby de-centred (Suchman 2004), with shifts and transformations entailed herein, through circulation and knowledge building (Nafus and Anderson 2009). Moreover, in the shaping process, spaces of interaction come into being. Here, seemingly contradictory or complementary interests and concerns, even contingently at times, juxtapose and consequently transform (Sørensen 2002). It is our argument that this underscores staging as sensitizing toward an actionable way forward, guiding and informing the collective shaping of design and innovation as a constructive endeavour.

Actions that lead to shaping are not delimited to that of the designer. And thus, while staging is very much one of the ways in which to intervene in the shaping, staging is thereby also not privileged to the designer alone. This indicates that staging entails means by which processes and outcomes may be attempted to be dealt with through political navigation, configuration of interactions, and more. Staging is hereby integral to qualifying and transforming, actively engaging, informing but also shaping emerging stakes and agendas. Shifting actors from pre-existing to new and emerging, come to shape and alternatively configure interactions in both process as well as outcomes. Staging becomes then not an a priori, prescriptive undertaking, but, rather, one which entails navigation through a dynamic terrain with multiple and equivocal agendas. This has bearing on knowledge creation that can thus be far from structured, well-planned, and acted out accordingly. It has been noted in innovation literature that coordination of knowledge and competencies in innovation calls for productively managing tensions and opportunities in boundary-spanning activities (Carlile 2002; Quintas 2005; Bathelt et al. 2017; Van de Ven and Zahra 2017). How agency unfolds to shape and make a difference through such a qualifying process is a strategic concern to design and innovation processes and their outcome(s).

Vinck and Tanferri (Chapter 15, this volume), drawing on the theatre metaphor, delving into and informed by actual theatrical practice. They emphasize staging as integral to setting up and rehearsing, through confronting, negotiating and stabilizing heterogeneous stakeholders and elements. Qualification and redefinition, and herein also iterations and rehearsals, are thus an integral aspect of the course of innovation. The productive character of analytically as well as strategically engaging the development of such meaning in the knowledge creation process becomes one crucial aspect of staging in innovation and design.

In addition to the process of stabilization, staging bears with it also the productive work of destabilization (for example in questioning preconceived notions and taken-for-grantedness, e.g. through de-familiarizing and leveraging ethnographic strangeness). We thus see the configuration of spaces, and the translation processes these entail as significant for understanding and problematizing the unravelling, destabilizing and subsequently stabilizing of new networks of relations.

STAGING AS THE CREATION OF SPACES ENABLING BOUNDARY CROSSING

The shaping of technology or the creation of something new compared to the pre-existing, or taken for granted, most often takes its point of departure from or is enabled in situations where technological, organizational or knowledge boundaries are being crossed (Clausen and Koch 2002; Russell and Williams 2002). Similarly, Bucciarelli (1994) has explored how innovation in design engineering involves the bridging of established engineering practices in the form of object worlds. Carlile (2002), in drawing upon the notion of boundary objects (Star and Griesemer 1989), has pointed to how the creation of new product solutions can be enabled through knowledge being transformed across social worlds and the knowledge practices entailed.

One of the main strategic concerns for the staging of design and innovation as presented in this book is how to create spaces where established practices and networks can be bridged. Bates and Juhl (Chapter 10, this volume) describe a case of how the transformation of research into a technology that can be commercialized is staged as a process to enable referential alignment across the differing criteria of research and innovation, seen as different forms of knowledge practices. In this context, staging becomes a strategy of setting up a test space that performs as a boundary object, across which the diverse criteria and methods from the world of research and technology are coordinated.

Another example of staging a space in order to bridge different worlds is presented in Chapter 7 (this volume). Here Clausen and Gunn present a case of staging a space set-up in order to mediate between the highly model-based engineering knowledge practices within the indoor climate, and user representations from ethnographic accounts based on field studies of the same. This space of knowledge mediation was staged as a series of workshops where engineers representing research and component manufacturers from the building sector carried out collective sensemaking and co-analysis of collected material representing user practices together with designers. See also Kjærsgard (2013) for a similar analysis.

What characterizes the staging of these two spaces are that they are being configured in specific ways. While these spaces enable collectives to negotiate and act together, they are also the outcome of negotiations of interests. Both cases concern attempts to cross boundaries where collaborations were not already being supported by institutionalized measures. The collaborations draw together a diversity of actors representing diverse and far from aligned knowledge practices and/or networks, where the actors staging and the participants agree in expecting a potential innovative outcome.

In both cases, we can also identify how the staging process includes arguments of problematization and interessement, revolving around discourses of innovation from a shared interest in sustainable industrial solutions (Bates and Juhl) or a common interest in a public funded program for user-driven innovation (Clausen and Gunn). There are common interests across the boundaries, for both cases, in creating something new. While this is so, the agency created through such spaces depends very much on the level of interest and commitments offered towards the shared space, versus the commitments shown towards the dominant 'home' practices. Staging of bridging spaces have to handle fragile boundaries and competing outside agendas. Here, the development of arguments that motivate participation is thus central (see also Hoffmann and Munthe-Kaas, Chapter 14, this volume).

The staging, here the purpose-oriented production, circulation and use of objects, play a key role in the framing of interactions and subsequent enabling of collective knowledge production and coordination. But, the two spaces elaborated in the respective chapters are staged to perform in diverse ways. Material knowledge objects in both cases are key in mediating interactions across diverse worlds and networks. In the test space (Bates and Juhl) an intermediary object (Vinck 2011), in the form of a test rig, was originally staged as a means to demonstrate the functionality and reliability of the new environmentally friendly lubricant. This came to serve as a boundary object as its position in the network changes, turning into a coordinating device across the criteria and interests of different organizational settings, while in the indoor climate case (Clausen and Gunn), intermediary objects in the form of representations of user practices were translated through mediating dialogues into provocative statements. This took place in the format of 'comfort themes' with the purpose of instigating changes in indoor climate conceptions upstream to the home organizations and research units of the participating engineers. The intermediary objects are not only key in the configuration of the shared space, but they can also be staged to perform particular work as part of a collective navigational process. The role of material objects in the staging of spaces evolving over longer stretches of development is further elaborated by Dorland (Chapter 6, this volume). Instead of focusing on the staging of singular bridging spaces, Dorland points at how objects are constructed to structure organizational spaces aligning with relevant stakeholders, and enabling collaboration across never ending iterations of organizing.

STAGING AS POLITICAL NAVIGATION AND ENACTMENT OF NETWORKS FOR INNOVATION

Several studies have pointed to the difficulties encountered by mature organizations wanting to support explorative innovative activities in a development organization geared towards exploitative and more routine innovation (Tushman and O'Reilly III 1996; Dougherty 2008; Jensen et al. 2018). Setting up a space for innovation or sustainable transition in an organizational context depends on the political support that can be obtained from internal and external actors and networks. For instance, in the case above presented by Bates and Juhl, financial and organizational resources were enrolled by navigating and enacting relevant business and sustainability strategies.

The staging of spaces for development in a mature organization is considered by Brønnum and Clausen (Chapter 9, this volume). The authors present a case on how a project manager and his team enabled explorative and radical front-end innovation in a highly institutionalized development organization, with a prevailing tradition for incremental innovation. In this case, staging becomes a strategy for navigating and (re-)enacting certain conditions of possibilities (Law 2002) in a particular way. Through a (re-)enactment of established conditions an alternative development agenda was made possible by strategically mobilizing actors from top management, reframing strategic elements and the meaning of the local development culture.

A similar case to such staging of the radical front end of innovation would be the current attempts to escaping carbon lock-ins by challenging established production systems and value chains transitioning towards a more sustainable future (Doganova and Karnøe 2012). Returning to contributions in this book, another mature organization's attempts to stage a transition, towards a circular economy is being addressed by Huulgaard et al. (Chapter 8, this volume). The intention was to set up temporary spaces for stimulating experimentation and learning across organizational stakeholders such as business divisions, development, technology, environmental concerns and potentially top management. The staging strategy involved negotiating a suitable and possible configuration and sequence of spaces based on an understanding of the company's internal and external contextual situations.

In both cases from this book, it can be seen how critical it is to gain support from actors, embedded in established networks such as strategic management and downstream development, or in business units deeply embedded in manufacturing practices and value chains. But the strategies for enrolling support were rather different in the two cases. In Brønnum and Clausen's case an internal project manager embedded in the organization staged a space for concept development as part of the development organization by enacting key conditions for development. Huulgaard et al., on the other hand, report on staging spaces for transition where participants from various parts of the organization were invited by a supportive branch (environmental department) to participate in bounded spaces outside the formal structures designed to escape the dominant logic and routine practices. The implication is that the distributed agency of the participating actors is focused and committed to a greater or lesser extent depending on the staging strategy and that staging strategies are highly situated and depend on the kind of space being staged.

Another key observation is the importance of continued political navigation of unfolding opportunities also triggered through realignment. Huulgaard et al. point at the role of negotiations before and between the planned spaces and in particular the political work of making sense of and relating to unfolding events which may open up, and shutting down opportunities. Staging is a political strategic endeavour that includes a multitude of staging moves (Pedersen 2020) and extends across the preparation and planning of events all of which require skilful political navigation (Dawson 2000). Brønnum and Clausen point at the constitutional aspects of this political navigation – as a way to identify the possibilities of re-enactment of established constitutional networks. Also, the sources of political navigational skills vary when compared across the two cases. An embedded internal project manager has a possibility to draw on political navigational experiences and a deep knowledge of the actors and objects to be re-enacted. In comparison, an external researcher, even when collaborating with an internal actor, has to invest time and effort together with the company collaborator in order to understand the organizational conditions of possibilities and politics, in attempting to intervene.

STAGING AS THE CIRCULATION OF OBJECTS THROUGH TRANSFORMATIVE SPACES

Participatory processes of design and innovation facilitate engagement of heterogeneous actors, artefacts and concerns in and across spaces, as they are leveraged and mutually transformed (Telier 2011). Authors of several chapters in this book open such capacities for transformation up to scrutiny. Sanders (Chapter 4, this volume) addresses engagements with objects in the process of co-design where staging can take on various levels of scale and pervasiveness. The role of designers and design objects is addressed by Pedersen and Brodersen (Chapters 3 and 5, this volume) where a more navigational role of the designer is emphasized. Bates' contribution (Chapter 11, this volume) takes up the workings of objects in engineering practice and in relation to the corporate organizational setting. While Bates' work cannot be relegated to design per se, it addresses staging in innovation through processes akin to design, where objects of engineering knowledge and practice serve to enable navigation and negotiation in wider collective processes of the organization. Central to the notion of staging in regards to the circulation of objects is how objects demonstrate equivocality and multiplicity, and enter into displacements with spaces, in reciprocal instances of circulation and unfolding. As intermediary objects they enable agency as a source of knowledge production, where neither the objects, nor the spaces are fixed and unchanging, giving the role of objects in this regard a political and navigational significance.

Pedersen and Brodersen's contributions showcase a common framework in between them, where productive tension is brought to light, in relation to the original design challenges at hand. Tension is made manifest through the concrete materialization and circulation of design objects by the designer(s). These serve toward the further staging of collectively transforming the spaces of negotiation with a shifting character of implicated actors. In one contribution (Chapter 5), particularly noteworthy is the materialization of an otherwise marginalized actor, the group of people with dementia in the nursing home. The case illustrates what was the lack of use of existing sensory stimulation technology by the elderly residents with dementia leading to gaining insight into what makes the residents' daily lives meaningful and less stressful.

The point of departure is how a team of design engineering students take on nursing home managers' interest in collaborating on devising how care givers as well as residents with dementia may benefit from sensory stimulation technology. The student design team navigate through the process with in-depth field work materialized as thick descriptions and interpretation of field insights. They further deploy a range of engineering design methods through a host of materialities and other modes of engagement in successive attempts to mediate between the residents and the design team, subsequently revisiting nursing home managers' concerns. Continuously working at reframing and thereby also qualifying concerns that may enter into negotiation spaces, the design team can be seen to stage the spaces of interaction and their transformation. The likes of affinity diagrams and design specifications in their evolving process, at the hands of the design team, underscore the transformative and far from fixed and stable character of these objects. These intermediary objects are rendered stable to allow for collective interaction, only to be rendered fluid (again) allowing for further transformation, and again rendered stable.

In this light, putting design objects into circulation, such as from fieldwork analyses and enrolling participants in the course of specification work (through prioritizing) and prototype testing iteratively help to translate and qualify what is at the core of an otherwise precarious process. Staging design through the circulation of design objects prompts the participants and designers alike to articulate multiple concerns and question how these come to be represented.

In Pedersen and Brodersen's other contribution, designers' engagement with the BoP (Bottom of the Pyramid) diabetes project (Chapter 3, this volume), entails an entirely different approach to the client company's heretofore reliance on the medical expertise of professionals in diabetes treatment that the Western market hinges on. Envisioning an innovation ecosystem is staged by the designers through design objects to enrol not only local medical expertise in India but also the lay-experiences and concerns of people living with diabetes in the households of city slums in New Delhi. The staging grounds insights and crucial user concerns as to quality in the provision of the medicine made accessible. Thus, design objects in the form of non-human participants (such as value cards and scenario cards) open up for matters of concern to be negotiated through a wider constituency, not expected in the client company's traditional focus. Their staging manoeuvre extends into how such insights are brought (back) into the organizational setting of the client company. First, the project manager is invited to partake of the findings and proposed solution for an insulin provision system based on BoP. The design objects of this presentation are through the input of the project manager further reworked and thus reconfigured for circulation as a formal presentation, in the company space.

In Bates' contribution to this volume (Chapter 11), the circulation of objects and the transformative capacity of spaces of interaction take centre stage in knowledge building and reframing of concerns. Engineering practice leverages objects such as models and simulations, which may be made to strategically challenge and occasion opportunities scoped and framed in multiple ways. With the backdrop of the organizational framework of the Technology Development Project (TDP) team and interests of the management, uncertainty as to customer interest manifest sources of tension. What is demonstrated is the significance of objects being potentially relevant to implicated networks be it management or engineering, in different ways, and how it becomes productive for the qualification of the hydraulic steering technology's transition into a product development project. Staging through objects transforms objects and knowledge relations in this regard by way of a destabilization and realignment through successive movements. In contrast to Pedersen and Brodersen's contributions, the objects of engineering practices are not design objects per se; nevertheless, objects such as Technological Readiness Assessments (TRAs) and the business case put into circulation achieve effects of reconfiguration central to staging.

The actors, together with the unfolding (potential and realized) relevance of the objects in circulation, undergo transformation also. They are in no way fixed. It gives the circulation of objects and the transformative capacity entailed in this regard political character. Staging innovation inherently deals with shifting alignments of objects as well as actors, and a significant aspect of navigation to deal with the contingencies and tensions this entails. This bears with it also the relevant question of discerning seeming tensions (controversy) as to the meaning of objects, and how these enter into negotiation and processes of stabilization through reconfiguration.

STAGING WITH EXPERIMENTS THROUGH CONFIGURING AND ARTICULATING IN URBAN PLANNING AND LIVING

Making tangible possibilities of alignment through enabling reprioritizations as to conceptions of health/wellbeing (Pedersen and Brodersen's chapters) or organizational agendas and interests at play (Bates' chapter) gives the staging of design and innovation political character. A set of contributions may be said to take on political undercurrents more explicitly, problematizing socio-politics in their framing of design and innovation processes and their staging. These are the contributions of Jensen (Chapter 12), Johansen and Lindegaard (Chapter 13) and Hoffmann and Munthe-Kaas (Chapter 14), which are works that open up to innovative experimental approaches, steering away from business-as-usual engagements. Whether it is to engage expertise in urban planning or consumers as practitioners in the context of Living Labs, these contributions touch upon the *pre*-configuration, highlighting inventive ways in which the field or subjects of study are engaged. This is of strategic interest to staging. Preconfiguration enters into the ways in which challenges are being framed, and how actors, materialities, practices and knowledge/ insights are being articulated. In this regard preconfiguration bears upon the inclusion and exclusion of issues. Yet, it is through translations and reconfiguration that strategic network-building founded upon a possible reframing and rearticulation of such heterogeneous elements can ensue.

For instance, the ENERGISE Living Lab elaborated upon by Jensen (Chapter 12, this volume), as a researcher-led intervention, broaches materials, meanings and competences guided by a practice theory perspective. It breaks away, also in terms of ontological underpinnings, from any pre-configuration as to traditional consumer behavioural framings of resource-intensive energy use in households in Denmark. As part of the project's undertakings, interventions in heating and laundry practices are specifically staged to grapple with ideas about, and material arrangements concerning domains of everyday practice. These interventions open up to otherwise preconfigured conceptions into existing demand for energy. Through experimenting and learning in relation to energy use, the interventions lend agency to households, to challenge and reconfigure these practices, and on this ground engage the laundry machine and building industry as well as municipalities.The different configuration and the concrete practices give the households agency to challenge and reconfigure these practices, in terms of such ideas and material arrangements.

Hoffmann and Munthe-Kaas (Chapter 14, this volume) highlight, through a progression of cases in the history of urban planning in specific geographical settings of Copenhagen, how (through field experiments) urban planning is opened up, highlighting also ontologically how interventions break away from the controlled and planned out to having bearing on stakeholder transformation and to the making of alternative futures (Ehn et al. 2014). Hoffmann and Munthe-Kaas refer to invitational devices in the interventions that configure knowledge and experience through controversy and collective democratic engagement. This is in stark contrast to previous urban planning initiatives, and particularly how such devices serve to configure participation in particular ways (Marres 2013).

Johansen and Lindegaard (Chapter 13, this volume) also take up preconfigured practices embedded institutionally, as relates to urban nature planning in the municipal administration of Copenhagen. The contribution specifically focuses on field interventions in the form of strategy workshops that challenge institutional participants from across the municipal organization in charge of urban nature planning. This is done through the material arrangements and cultures/epistemic underpinnings that participants are prompted to explore and articulate collectively. The initial workshop faces opposition from the participants, as it is staged to collaboratively devise a 'stronger' urban nature management tool. The participant positions are leveraged from such a counterproductive stance toward taking on an explorative and constructive character in the situated intervention in what ensues, taking note of the management orientation's discontent. The workshop's initially preconfigured agenda is redressed to enable new articulations and configurations in the work with urban nature by and among the participants in the municipal administration.

These contributions bear upon them the experimental character of engagements in the 'field'. This is in regards to the uncertainty of the unfolding and outcomes of the process(es), in keeping with recent scholarly work on fieldwork devices, where heterogeneous, material, social and epistemic aspects are underscored, in the creative unfolding and co-production collective field engagements (Estalella and Criado 2018). What is more, the three contributions address pre-configuration as a significant aspect that staging through experimental approaches also grapples with. Whether it be through researcher-led interventions into energy practices, opening up urban planning to democratic exchange or articulating municipal endeavours with urban nature, staging thus also highlights strategic aspects to the inclusion and exclusion that might be preconfigured, and not in the least configuring, network-building.

CONCLUDING PERSPECTIVES

The present book addresses staging at the very crux of design and innovation. At the core of societal development and processes of societal change the imperative to innovate brings with it the challenge of multiple and often also seemingly contradictory concerns. Seemingly innovative prospects are wrought with uncertainties and with tension, where design may be construed as playing an integral part in unravelling and synthesizing such issues. Staging design and innovation may open up to ways in which to deal with uncertainties as well as reconfiguring preconceived ideas and practices in constructive ways to make a productive difference in the shaping of outcomes.

What has become clear from reading across the chapters of this book is that staging is applied to a variety of problems and issues as well as the prospects for solving them. In the previous sections we have drawn up and elaborated on concrete strategies for staging which exemplify, substantiate and even expand the repertoire of staging actions mentioned in Chapter 2 (this volume). These strategies demonstrate how staging can be applied to strategically bridge established networks and mediate across highly diverse knowledge practices in order to seek transitioning systems of knowledge or new innovative possibilities. Another strategy is concerned with the political navigation and enactment of commitment of agency to emerging spaces for front-end innovation and transitions from established networks. A third strategy demonstrates how a deliberate development and circulation of objects through transformative spaces can prompt articulation and representation of concerns to enable the development of collaborative agency, but also that staging innovation has to deal with shifting alignments of objects as well as actors. Finally, strategies of staging as experimentation demonstrate how preconceived ideas of urban development, urban nature and living practices can be challenged. These are means of constructively dealing with challenges, possibilities and even differing ontologies that staging strategies aiming to bridge differences and reconfigure spaces and objects can draw on.

From these strategic perspectives on what staging may offer to shape design and innovation, we have pointed to political navigation as a key aspect. The positions of various actors – both those taking on staging roles and those attempting to become enrolled – and the spaces they may be enrolled in are not fixed but only provisional. They often escape preconceived understandings and simple identification and are subject to change as staging unfolds influenced by competing perspectives and network memberships. Consequently, as we have seen, actor-positions and spaces can be envisaged as being in tension with one another. Constructive engagements with such tensions are addressed throughout the contributions and point at fruitful processes of collective evaluation and continuous production of new knowledge, understandings and configurations. The skilful navigation of these tensions may be helped by notions such as negotiation spaces (Pedersen 2020) and intermediary objects (Vinck 2011).

Staging offers strategies for opening up preconfigured notions and taken for granted ontologies. But what ontologies to question and which to build upon, or which type of concerns to attend to? Setting up negotiation spaces (Pedersen 2020) for instance include considerations concerning who and what to include and not to include and when and in what way. Here a navigation of the extended scenery or development arena (Jørgensen and Sørensen 2002) could be considered. In the development arena, we may discover multiple networks – that is to say multiple ontologies – and actors with ambitions and strategies to stage design or innovation or maintain the already established from a competing perspective. To continue the lessons learned from the cases presented in this book, we would suggest a further qualification of the political navigational strategies employed by way of analysing and understanding unfolding conditions of possibilities in the wider development arena. What work do we expect the intermediary objects or the emerging concepts to perform as they are circulated and being subject to deliberate framing and reframing and while they translate and are being translated as they engage diverse networks? By taking into account the already existing actor worlds (Callon 1986; Jørgensen 2012; Heiskanen et al. 2018) or networks in the making, their narratives, strategies, assumptions, framing objects and concerns, more informed choices could be made about how to enter the arena, setting up spaces, and circulating objects.

Professionals engaged in the staging of design and innovation have to deal with a multitude of existing and emerging interests that hardly present themselves in any orderly fashion, at times competing or at the very least, convoluting. They are subject to changing and unclear agendas and to shifting concerns, and, in this context, the development of navigational skills for the staging professional is highly recommended, if not indispensable. Here, the practitioner can draw on lessons learned through the continuing rehearsals that Vinck and Tanferri (Chapter 15, this volume) point to as inherent in a staging effort. We suggest that rehearsals and continuing design inquiries such as in spirals of 'appreciating, reframing, experimenting and reappreciating' (Telier 2011: 85) include navigational reflections (Broberg and Hermund 2004) and ongoing explorations aimed at probing into and sensing future opportunities. In this way we propose to advance staging as a sensitizing concept towards the interactional aspects of design and innovation in the hands of professionals.

This book presents a collection of cases from practice that the professionals and their educators may learn from and draw on. Together, they offer a repertoire of actionable strategies and concepts for reflection. On these grounds we can suggest that professionals develop their navigational staging skills through rehearsals, design inquiries and ongoing negotiations attending to diverse articulations of interest and the ever-changing conditions of development.

REFERENCES

Bathelt, H., P. Cohendet, S. Henn and L. Simon (2017), 'Innovation and knowledge creation: Challenges to the field', in H. Bathelt, P. Cohendet, S. Henn and L. Simon (eds), *The Elgar Companion to Innovation and Knowledge Creation*, Cheltenham, UK and Northampton, MA, USA: Edward Elgar Publishing, pp. 1–21.

Broberg, O. and I. Hermund (2004), 'The OHS consultant as a "political reflective navigator" in technological change processes', *International Journal of Industrial Ergonomics*, **33**, 315–326.

Bucciarelli, L. (1994), *Designing Engineers*, Cambridge, MA: The MIT Press.

Callon, M. (1986), 'The sociology of an actor-network: The case of the electric vehicle', in M. Callon, J. Law and A. Rip (eds), *Mapping the Dynamics of Science and Technology*, London: Macmillan, pp. 19–34.

Carlile, P. (2002), 'A pragmatic view of knowledge and boundaries: Boundary objects in new product development', *Organization Science*, **13** (4), 442–55.

Clausen C. and C. Koch (2002), 'Spaces and occasions in the social shaping of information technologies', in K.H. Sørensen and R. Williams (eds), *Shaping Technology, Guiding Policy: Concepts, Spaces and Tools.* Cheltenham, UK and Northampton, MA, USA: Edward Elgar Publishing, pp. 223–48.

Dawson, P. (2000), 'Technology, work restructuring and the orchestration of a rational narrative in the pursuit of "management objectives": The political process of plant-level change', *Technology Analysis and Strategic Management*, **12** (1), 39–58.

Doganova, L. and P. Karnøe (2012), *The Innovator's Struggle to Assemble Environmental Concerns to Economic Worth: Report to Grundfos New Business, March 2012*, Bjerringbro: Grundfos New Business, accessed 5 August 2020 at https://www.industriensfond.dk/sites/default/files/grundfos.pdf.

Dougherty, D. (2008), 'Managing the "unmanageables" of sustained product innovation', in S. Shane (ed.), *Handbook of Technology and Innovation Management,* Hoboken: John Wiley and Sons, pp. 173–93.

Ehn, P., E.M. Nilsson and R. Topgaard (eds) (2014), *Making Futures: Marginal Notes on Innovation, Design, and Democracy*, Cambridge, MA: The MIT Press.

Estalella, A. and T.S. Criado (eds) (2018), *Experimental Collaborations: Ethnography through Fieldwork Devices*, New York and Oxford: Berghahn.

Heiskanen, E., E-L. Apajalahti, K. Matschoss and R. Lovio (2018), 'Incumbent energy companies navigating the energy transitions: Strategic action or bricolage?', *Environmental Innovation and Societal Transitions*, **28**, 57–69.

Jensen, A.R.V., C. Clausen and L. Gish (2018), 'Three perspectives on managing front end innovation: Process, knowledge, and translation', *International Journal of Innovation Management*, **22** (7), accessed 17 October 2018 at https://doi.org/10.1142/S1363919618500603.

Jørgensen, U. (2012), 'Mapping and navigating transitions – The muliti-level perspective compared with arenas of development', *Research Policy*, **41**, 996–1010.

Jørgensen, U. and O. Sørensen (2002), 'Arenas of development: A space populated by actor worlds, artefacts and surprises', in K.H. Sørensen and R. Williams (eds), *Social Shaping of Technology: Concepts, Spaces and Tools*, Cheltenham, UK and Northampton, MA, USA: Edward Elgar Publishing, pp. 197–222.

Kjærsgaard, M.G. (2013), '(Trans)forming knowledge and design concepts in the design workshop', in W. Gunn, T. Otto and R.C. Smith (eds), *Design Anthropology: Theory and Practice*, London and New York: Bloomsbury Academic, pp. 51–67.

Law, J. (2002), 'Objects and spaces', *Theory, Culture & Society,* **19** (5/6), 91–105.

Marres, N. (2013), 'Why political ontology must be experimentalized: On eco-show homes as devices of participation', *Social Studies of Science* **43** (3), pp. 417–43.

Nafus, D. and K. Anderson (2009), 'Writing on walls: The materiality of social memory in corporate research', in M. Cefkin (ed.), *Ethnography and the Corporate Encounter: Reflections on Research in and of Corporations*, New York and Oxford: Berghan, pp. 137–57.

Pedersen, S. (2020), 'Staging negotiation spaces: A co-design framework', *Design Studies*, DOI: 10.1016/j.destud.2020.02.002.

Quintas, P. (2005), 'Managing knowledge and innovation across boundaries', in S. Little and T. Ray (eds), *Managing Knowledge: An Essential Reader*, London and Thousand Oaks: Sage Publications, pp. 255–71.

Russell, S. and R. Williams (2002), 'Social shaping of technology: Frameworks, findings and implications for policy', in K.H. Sørensen and R. Williams (eds), *Shaping Technology, Guiding Policy: Concepts, Spaces and Tools*, Cheltenham, UK and Northampton, MA, USA: Edward Elgar Publishing, pp. 37–131.

Sørensen, K.H. (2002), 'Social shaping on the move? On the policy relevance of the social shaping of technology perspective', in K.H. Sørensen and R. Williams (eds), *Shaping Technology, Guiding Policy: Concepts, Spaces and Tools*, Cheltenham, UK and Northampton, MA, USA: Edward Elgar Publishing, pp. 19–35.

Star, S.L. and J.R. Griesemer (1989), 'Institutional ecology, "translations" and boundary objects: Amateurs and professionals in Berkeley's Museum of Vertebrate Zoology, 1907–39', *Social Studies of Science,* **19** (3), 387–420.

Suchman, L. (2004), 'Decentering the manager/designer', in R.J. Boland Jr and F. Collopy (eds), *Managing as Designing*, Stanford, CA: Stanford Business Books, pp. 169–73.

Telier, A. (T. Binder, G. De Michelis, P. Ehn, G. Jacucci, P. Linde and I. Wagner) (2011), *Design Things*, Cambridge, MA and London: The MIT Press.

Tushman, M.L. and C.A. Reilly III (1996), 'Ambidextrous organizations: Managing evolutionary and revolutionary change', *California Management Review,* **38** (4), 8–30.

Van de Ven, A. and S.A. Zahra (2017), 'Boundary spanning, boundary objects, and innovation', in F. Telt, C. Berggren, S. Brusoni and A. Van de Ven (eds), *Managing Knowledge Integration Across Boundaries*, Oxford: Oxford University Press, pp. 241–54.
Vinck, D. (2011), 'Taking intermediary objects and equipping work into account in the study of engineering practices', *Engineering Studies* **3** (1), 25–44. DOI: 10.1080/19378629.2010.547989.

Index